AF560056

THE ARCHAIC AND THE EXOTIC

THE ARCHAIC AND THE EXOTIC

Studies in the History of Indian Astronomical Instruments

Sreeramula Rajeswara Sarma

MANOHAR
2008

First published 2008

ISBN 81-7304-571-2

Published by

Ajay Kumar Jain for
Manohar Publishers & Distributors
4753/23 Ansari Road, Daryaganj
New Delhi 110 002

Typeset at

Digigrafics
New Delhi 110 049

Printed at

Lordson Publishers Pvt Ltd
Delhi 110 007

In memory of

my mother

Sreeramula Sakunthalamma

and

my aunt

Kaluvalapalli Rajalakshmi

Contents

List of Illustrations 9

Preface 13

I. THE CONTEXT

1. Indian Astronomical and Time-Measuring Instruments: A Catalogue in Preparation 19
2. Astronomical Instruments in Brahmagupta's Brāhmasphuṭasiddhānta 47
3. Perpetual Motion Machines and their Design in Ancient India 64
4. Astronomical Instruments in Mughal Miniatures 76

II. THE WATER CLOCK

5. The Bowl that Sinks and Tells Time 125
6. Announcing Time: The Unique Method at Hayatnagar, 1676 136
7. Measuring Time with Long Syllables: Bhāskara I's Commentary on Āryabhaṭīya, Kālakriyāpāda 2 143
8. Setting up the Water Clock for Telling the Time of Marriage 147

III. THE ASTROLABE

9. Sulṭān, Sūri and the Astrolabe 179
10. The Lahore Family of Astrolabists and their Ouvrage 199

11. The Ṣafīḥa Zarqāliyya in India 223

12. Yantrarāja: The Astrolabe in Sanskrit 240

13. Kaṭapayādi Notation on a Sanskrit Astrolabe 257

IV. THE CELESTIAL GLOBE

14. From al-Kura to Bhagola: On the Dissemination of the Celestial Globe in India 275

15. Two Mughal Celestial Globes 294

Index 309

Illustrations

1.1 Water clock made of coconut shell, Government Museum, Madras 29
1.2 Vertical gnomon on a horizontal dial, Firdaus Manzil, Patna 32
1.3 Carved wooden column dial, a composite picture 34
1.4 Indo-Persian quadrant, National Maritime Museum, Greenwich 34
1.5 *Dhruvabhrama-yantra*, Jai Singh's Observatory, Jaipur 35
1.6 Astrolabe by 'Īsà ibn Allāhdād Asṭurlābī Humāyūnī Lāhūrī 38
1.7 Sanskrit astrolabe, single plate for the latitude 22;35,39°N, dated 1941 VS/AD 1884 41
1.8 Celestial globe by Bahlūmal Lāhūrī, dated 1258 AH/AD 1842, The Science Museum, London 43
1.9 The astrolabist family of Lahore and their instruments 44
3.1 Perpetual motion wheel according to Brahmagupta 65
3.2 Bhāskara's first model of perpetual motion wheel 66
3.3 Bhāskara's second model of perpetual motion wheel 67
3.4 Perpetual motion wheel from an Arabic manuscript 67
4.1 "Birth of Murad" by Bhurah and Baswan from the *Akbarnāma*, Victoria & Albert Museum, London 109
4.1A "Astrologers casting the horoscope", detail from Fig. 4.1 110
4.2 "Astrologers explaining the horoscope to the king", from the *Akbarnāma*, British Library, London 111
4.3 "Astrologers casting the horoscope", Museum of Fine Arts, Boston 112
4.4 "Astrologers casting the horoscope", Chester Beatty Collection, Dublin 113
4.5 "Astrologers casting the horoscope". British Library, London 114

4.6 "Jahāngīr seated on an hourglass", by Bichitr, Freer Gallery of Art, Washington 115
4.7 "Angels holding a crown, celestial globe and ring dial", Sackler Gallery of Art, Washington 116
4.8 "The Astrologer" from the Jahāngīr's Album, Náprstek Museum, Prague 117
4.9 "Astronomer with his disciples", Shāh Jahān's Album, Musée Guimet, Paris 118
4.10 "Bazar astrologer with his clients", from the *Akhlāq-i Naṣīrī*, Museum Rietberg, Zürich 119
4.11 "Noah's Ark", Freer Gallery of Art, Washington 120
4.12 Two *Cūḍāyantras* from Jai Singh's Observatory, Jaipur 121
4.13 The larger *Cūḍāyantra* from Jai Singh's Observatory, Jaipur 121
5.1 Working principle of the sinking bowl type of water clock 126
5.2 *Ghaṭī-yantras* at the Pitt Rivers Museum, Oxford 130
5.3 Emperor's time piece from Burma 131
10.1 Front of the astrolabe dated 1031/1621 by Muḥammad Muqīm 210
10.2 Back of the astrolabe dated 1031/1621 by Muḥammad Muqīm 211
11.1 The Zarqālī astrolabe by Ḍiyā' al-Dīn. Front view 228
11.2 Front of the astrolabe. Detail from the upper left quadrant 229
11.3 Back view of the astrolabe 231
11.4 Back of the astrolabe. Detail showing the inscription on the crown, and the plaque 233
12.1 Rāghavajit's Sanskrit astrolabe with five tympans, AD 1668 253
12.2 A Sanskrit astrolabe with a single tympan for the latitude of 27° N 254
12.3 Sanskrit astrolabe manufactured by Śivalāla in 1879 for the latitude of Bundi at 25;30° N 255
13.1 Sanskrit University astrolabe; front view with the star map 262
13.2 Sanskrit University astrolabe; back view; with the alidade, nut and pin 265

13.3 Detail of the plate for latitude 18° 270
13.4 Detail of the plate for latitude 28° Vamśāvalīnagara 270
14.1 Aligarh globe with stand 283
14.2 Aligarh globe: horizontal frame 284
14.3 Aligarh globe: plug near the South Pole 284
14.4 Aligarh globe: ecliptic and equator 285
14.5 Aligarh globe: maker's signature 285
14.6 Astronomer-philosopher surrounded by astronomical instruments 287
14.7 Angels holding the crown, celestial globe and ring dial 287
14.8 Sanskrit celestial globe from Jaipur 290
15.1 Globe A with stand 300
15.2 Globe A with the horizon ring 301
15.3 Globe B with stand. Constellations of the northern hemisphere 304
15.4 Globe B with stand. Constellations of the southern hemisphere 305
15.5 Globe A, detail. Constellation Argo Navis 306
15.6 Globe B, detail. Constellation Argo Navis 306

Preface

The papers collected in this volume are related to my investigations into the history of astronomical and time-measuring instruments in India. This history needs to be reconstructed from literary sources as well as from actual specimens that are extant. The first paper in this volume dwells on the importance of cataloguing the extant specimens of pre-modern astronomical instruments produced in India and gives an overview of those preserved in various museums in India and outside. Among the literary sources, Brahmagupta's *Brāhmasphuṭasiddhānta* offers the first systematic discussion of a large number of instruments and their use. The twenty-second chapter of this work which is exclusively devoted to instruments is analysed in the second paper. Questions of transmission from one culture area to another are discussed in the third paper entitled 'Perpetual Motion Machines and their Design in Ancient India.' Besides literary sources and actual specimens, there is yet a third source, viz., paintings. While looking at the astronomical instruments depicted in the Mughal miniatures, the fourth paper addresses the larger issue of exchanges between Sanskritic and Islamic traditions of astronomy and astronomical instrumentation. These four papers thus define the context in which my studies are undertaken.

The history of astronomical instrumentation in India is dominated by two mutually contradictory—yet complementary—currents: on the one hand the resilience of certain archaic instruments that held sway for long even after they had become obsolete; on the other, Indian astronomers' receptivity to exotic instruments from other cultures. Hence the title of volume: *The Archaic and the Exotic*.

The archaic is the sinking bowl variety of water clock called *Ghaṭī-yantra*, from which *ghaḍī*, the name for watches and clocks in many modern Indian languages, is derived. This type of water clock is mentioned for the first time in a commentary written in Sri

Lanka by Buddhaghoṣa in the first half of the fifth century and has been widely in use in all the Indian Ocean countries up to the beginning of the twentieth century. In India, there developed an interesting ritual about the setting up of this device for determining the auspicious moment for weddings, and a poignant legend grew about its non-function and the tragic consequences thereof. Even now the *Ghaṭī-yantra* is employed in certain places of worship of the Hindus, Jainas and Muslims. The horological vocabulary in almost all Indian languages is still reminiscent of the age-old practice of measuring time with this sinking bowl and broadcasting it by striking on a metal gong. Four papers in part II deal with these and other aspects of the water clock.

The astrolabe and the celestial globe are the exotic instruments received enthusiastically in India from the Islamic World. Of these two, the astrolabe is the more versatile instrument. It was held in high esteem in all the cultures from Spain to India. Though it may have been introduced into India even earlier, its study received an impetus in the second half of the fourteenth century through the interest of Fīrūz Shāh Tughluq. Under his auspices, astrolabe manufacture commenced at Delhi, and manuals were composed on its construction and use in Persian and in Sanskrit. The production of astrolabes reached its technical and artistic pinnacle in the Mughal period. Today there survive a very large number of exquisitely crafted Mughal astrolabes in several collections all over the world.

The Jaina monk Mahendra Sūri, who composed the first ever manual in Sanskrit on the astrolabe at the court of Fīrūz Shāh Tughluq at Delhi in 1370, was so impressed by the astrolabe that he called it *Yantra-rāja* ("king of the astronomical instruments") in Sanskrit. From his time up to the end of the eighteenth century, at least a dozen Sanskrit manuals were composed on the astrolabe. Simultaneously astrolabes were also produced with Sanskrit labels and scales. Today, Indo-Persian astrolabes with Arabic/Persian legends and Sanskrit astrolabes form the bulk of extant Indian astronomical instruments. Five papers in part III deal with the history of the astrolabe in India: its promotion by Fīrūz Shāh Tughluq in the fourteenth century, the dominant role played by a family of instrument makers from Lahore in the production of astrolabes under the patronage of Mughal rulers from Humāyūn to

Aurangzeb, Sanskrit manuals composed on the astrolabe, and certain individual specimens of Indo-Persian and Sanskrit astrolabes.

India introduced an important innovation in the production of the celestial globe at the close of the sixteenth century. Prior to this time, globes were made first as two hollow hemispheres and then joined together. Qā'im Muḥammad, an instrument maker from Lahore, adopted the ancient Indian technique of making metal images by the *cire perdue* process and cast the celestial globes in one piece. The history of the celestial globe in India is dealt with in the last two papers where I introduce, among others, an exceptionally fine celestial globe crafted by Ḍiyā' al-Dīn Muḥammad in 1653. Preserved at Aligarh Muslim University, it became the starting point of my survey of Indian astronomical instruments.

For this survey, I visited, during the last decade and a half, more than a hundred museums and private collections in India, Europe and North America, where I identified and studied some 430 specimens of pre-modern Indian astronomical instruments. I am currently preparing a descriptive catalogue of these instruments. The catalogue will contain historical surveys of each instrument-type, its use and geographical spread, besides full technical description of each specimen, with art-historical notes and photographic documentation. I shall also discuss the relation between the theoretical prescriptions on instrument making in Sanskrit astronomical treatises and their actual execution; and the interplay between the Sanskritic and Islamic traditions of instrumentation. I hope the present volume of studies will serve as a companion to the forthcoming Descriptive Catalogue.

It is my pleasant duty to thank the editors of the Journals and Felicitation Volumes for publishing these papers and for giving permission now to reprint them in this volume. Here the papers are reprinted as they originally appeared in publication with different referencing style; hence some divergence will be found here in the style of citations.

In the course of my studies on astronomical instruments, I was immensely benefited by the writings of several scholars; others facilitated my access to instruments in museums and private collections; yet others offered scholarly advice and much needed encouragement. For their invaluable help, I express my deep sense of gratitude to all these heads of museums, scholars, colleagues and friends: Dr R.G.W. Anderson (formerly Director of the British Museum, London, and now Visiting Fellow, Corpus Christi College, Cambridge); Prof. S.M. Razaullah Ansari (President, Commission for History of Ancient and Medieval Astronomy, International Union of History and Philosophy of Science, Aligarh); Dr A.K. Bag (Editor, *Indian Journal of the History of Science*, and Advisor, History of Science Programme, Indian National Science Academy, New Delhi); Prof. Nalini Balbir (Mondes iranien et indien, Université Paris III Sorbonne Nouvelle, Paris); Dr Jim Bennett (Director, Museum of the History of Science, Oxford); Prof. Owen Gingerich (Harvard-Smithsonian Center for Astrophysics, Cambridge, Mass); Prof. David King (formerly Director of the Institut für Geschichte der Naturwissenschaften, Johann Wolfgang Goethe Universität, Frankfurt); the late Mr Francis Maddison (Curator of the Museum of the History of Science, Oxford); the late Prof. David Pingree (University Professor of History of Mathematics and Classics, Brown University, Providence); Dr A.K.V.S. Reddy (formerly Director, Salar Jung Museum, Hyderabad, now Director General, National Museum, New Delhi); Prof. Emilie Savage-Smith (Professor of the History of Islamic Science, The Oriental Institute, University of Oxford); Prof. Fuat Sezgin (Director of the Institut für Geschichte der Arabisch-Islamischen Wissenschaften, Johann Wolfgang Goethe Universität, Frankfurt); Pt. Om Prakash Sharma (Superintendent, Jantar Mantar Observatory, Jaipur); Dr W.H. Siddiqi (Officer on Special Duty, Rampur Raza Library, Rampur); Mr Anthony J. Turner (President, Societè internationalè de l'astrolabe, Le Mesnil-le-Roi); Prof. G.L 'E Turner (formerly of Imperial College, London); Dr Kapila Vatsyayan (President, Indira Gandhi National Centre for the Arts Trust, New Delhi); Prof. Michio Yano (Dean, Faculty of Cultural Studies, Kyoto Sangyo University, Kyoto).

Finally, I must thank Mr Ramesh Jain for agreeing to publish this volume under the imprint of Manohar; he and his staff deserve all praise for their unfailing courtesy and for the elegant production of the volume.

SREERAMULA RAJESWARA SARMA

PART I

The Context

1

Indian Astronomical and Time-Measuring Instruments: A Catalogue in Preparation

0.1 Like literary documents, scientific instruments also constitute a valuable source for the history of science and technology of a people. In India, from the earliest times up to the beginning of the present century, various kinds of instruments have been in use for measuring time, or for taking the altitude and azimuth of heavenly bodies, or for visually demonstrating the apparent motion of the heavens. The construction and use of these pre-modern scientific instruments are described in several Sanskrit texts. A number of actual instruments are also extant in various collections in India and abroad. While a few Sanskrit texts have been published,[1] the instruments themselves did not receive much scholarly attention.[2] A disproportionately large corpus of literature exists no doubt on the gigantic masonry instruments erected by Sawai Jai Singh in the early eighteenth century, but hardly anything on their antecedents.[3]

First published in: *Indian Journal of History of Science*, 29.4 (1994) 507-28.

1. Mahendra Sūri, *Yantrarāja*, and Viśrāma, *Yantraśiromaṇi*, ed. K.K. Raikva, Bombay 1936; *Yantrarājaracanā of Jayasiṃha*, ed. Kedāranātha Jyotirvid, Jaipur 1953; *Yantrarāja-vicāra-viṃśādhyāyī of Nayanasukhopādhyāya*, ed. Bibhutibhushan Bhattacharya, Varanasi 1979; *Pratodayantra by Gaṇeśa Daivajña*, ed. Shakti Dhara Sharma, Kurali 1982. On the theory of observation, see Sarma, K.V., *Observational Astronomy in India*, Calicut 1990, and Ohashi, Yukio, Development of Astronomical Observation in Vedic and Post-Vedic India, *Indian Journal of History of Science*, 28.3, 185-251, 1993.
2. See 0.4 below.
3. For an exhaustive bibliography on Jai Singh, see Sarma, Sreeramula Rajeswara (ed & tr), *The Yantraprakāra of Sawai Jai Singh*, supplement to *Studies in History of Medicine and Science*, 9-10, 122-131, 1986-87.

Indeed not much was known about the portable metal instruments designed or collected by Jai Singh and his son Madho Singh, until I catalogued them in 1991.[4] But these extant instruments are a valuable part of our cultural heritage.

In the first quarter of the seventh century, Brahmagupta devoted an entire chapter of his *Brāhmasphuṭasiddhānta* to the instruments. In this chapter called appropriately *yantrādhyāya*, he gave the first systematic account of a large number of instruments, namely *ghaṭikā, śaṅku, cakra, dhanuṣ, turyagola, yaṣṭi, pīṭha, kapāla* and *kartarī*.[5] Following his example, subsequent astronomers like Lalla in the eighth century, Śrīpati in the eleventh and Bhāskara in the twelfth, adopted the same instruments and observational and computational techniques. Now the question arises whether these instrument types were ever actually manufactured and used in observation, or whether they were mere designs whose execution was never tested in practice, as in the case of the automata described in several texts.

Of course, even the design per se can be important in the history of ideas. Lynn White had argued that the Indian concept of the perpetual motion machine eventually led to modern power technology,[6] and the present writer has shown that the first perpetuum mobile was designed by Brahmagupta in the seventh century and not by Bhāskara in the twelfth, as Lynn White had

4. Cf. Sarma, Sreeramula Rajeswara, Portable Instruments at the Jaipur Observatory, paper read at the International Symposium on Indian and Other Asiatic Astronomies, Hyderabad-Jaipur 1991; idem, Manufacture of Portable Astronomical Instruments at Jaipur and the Contribution of Sawai Madho Singh, read at the XIX International Congress of History of Science, Zaragoza 1993.
5. Cf. Sarma, Sreeramula Rajeswara, Astronomical Instruments in Brahmagupta's Brāhmasphuṭasiddhānta, *Indian Historical Review*, 13, 63-74, 1986-87; reprinted in this Volume, pp. 47-63.
6. White, Lynn, Tibet, India, and Malaya as Sources of Western Medieval Technology, *American Historical Review*, 65, 515-526, 1960; reprinted in: idem, *Medieval Religion and Technology: Collected Essays*, Berkeley 1978, pp. 43-57; see also idem, *Medieval Technology and Social Change*, London 1964, pp. 129-151.

supposed.[7] Likewise, on the basis of the Indian method of determining the cardinal points (*diksādhanā*), the Arabs developed an instrument for finding out the meridian and the qibla which they called the "Indian Circle".[8]

In the reverse direction, contact with Islamic astronomy led, from about the fourteenth century onwards, to the composition of texts exclusively devoted to instruments. The earliest work of this nature is the *Yantrarāja* written by the Jaina monk Mahendra Sūri at the court of Fīrūz Shāh Tughluq in Delhi in 1370. The composition of all these texts must have been accompanied by the manufacture of actual instruments.

For a proper understanding, then, of the role played by the instruments in the development of scientific thought in India, the texts on instruments have to be studied together with the actual specimens. Sanskrit astronomical texts, like all śāstric texts, are notoriously brief. Even a lengthy compendium like Rāmacandra Vājapeyin's *Yantraprakāśa*, written in 1428 with an elaborate auto-commentary,[9] stands nowhere near the Arabic treatises on instrument-making with their precise and practical directions such as Mu'ayyad al-Dīn al 'Urḍī's account of the instruments he made for the Marāgha Observatory towards the middle of the thirteenth century,[10] or al-Jazarī's *Book of Ingenious Devices.*[11] Again, Sanskrit texts give only a general description of the instrument, but the variety in its execution and the geographical extent of its spread can only be known from the study of actual instruments.

Thus, on the one hand, actual specimens may help in understanding the often too brief texts. Conversely, textual knowledge

7. Sarma, Sreeramula Rajeswara, Perpetual Motion Machines and Their Design in Ancient India, *PHYSIS: Revista Internationale di Storia della Scienza,* 29.3, 665-676, 1992; reprinted in this Volume, pp. 64-75.
8. Wiedemann, Eilhard, Ueber den indischen Kreis, *Mitteilungen zur Geschichte der Medizin und der Naturwissenschaften,* 11, 252-255, 1912.
9. Cf. Sarma, Sreeramula Rajeswara, Rāmacandra Vājapeyin's Contribution to Mathematics and Astronomy, paper read at the VIII World Sanskrit Conference, Vienna 1990.
10. Cf. Seemann, Hugo J., *Die Instrumente der Sternwarte zu Marāgha nach den Mitteilungen von Al 'Urḍī,* Erlangen 1928.
11. Hill, Donald R. (tr), *The Book of Knowledge of Ingenious Mechanical Devices by Ibn al-Razzāz al Jazarī,* Dordrecht 1974.

helps in identifying an instrument and in dating its original design. Finally, this combined approach will throw better light on transcultural exchanges, especially between the Hindu and Islamic traditions of instrumentation during the medieval period.

0.2 I may illustrate this point with an example. Some time ago I was tantalized by a Mughal portrait of a roadside astrologer, which shows him holding a circular hoop-like object in his hand.[12] For those familiar with European astronomical instruments, it is not difficult to identify this hoop as the ring dial,[13] although of a larger kind. But why show it in a Mughal miniature painting, where one would normally expect an astrolabe, the instrument *par excellence* of the Islamic World? The same ring dial appears again and again in several other Mughal miniatures, which depict the birth of a royal personage, and the astrologers casting his horoscope.

The ring dial was common in Europe, especially in southern Germany and Austria where it was called *Bauernring* "farmers' ring", but it was not known to the Islamic World. Did it then obtain in India? The answer was soon furnished by Sawai Jai Singh's *Yantraprakāra* which I was then editing. It contains a long description of the ring dial under the name *Cūḍā-yantra,* together with a set of tables to be used for the latitude of Delhi.[14] About the same time, I also heard of the existence of two actual specimens in the stores of the Jaipur Observatory. Attached to one of them is a tablet on which the name *Cūḍā-yantra* was inscribed. Finally, an investigation into its antecedents showed that it was known to many earlier astronomers from Āryabhaṭa onwards. Varāhamihira calls it *Valaya-yantra*. In his *Yantraprakāśa* composed in 1428, Rāmacandra discusses three variants, one of which he designates as *Cūḍā-yantra*. Therefore, what the roadside astrologer in the

12. Cf. Sarma, Sreeramula Rajeswara, Astronomical Instruments in Mughal Miniatures, *Studien zur Indologie und Iranistik,* 16-17, 235-276, 1992. The painting is reproduced on p. 271 as Pl. 8; reprinted in this Volume, pp. 76-121; see p. 117, Fig. 4.8.
13. See Price, Derek, Precision Instruments, in: Charles Singer (ed), *A History of Technology,* vol. III, Oxford 1957, pp. 582-619, esp. p. 598, Fig. 351.
14. *Yantraprakāra of Sawai Jai Singh* (see n. 3 above), pp. 27-28, 79-82, 101, 105-114.

Mughal miniature held in his hand was not the European ring dial but the traditional Indian *Cūḍā-yantra*.

Thus, here we have a unique case in the history of scientific instrumentation where Sanskrit texts, Mughal miniature paintings and actual specimens from Sawai Jai Singh's observatory collectively document the history of the ring dial and its popularity through some twelve centuries from Āryabhaṭa to Jai Singh.

0.3 The study of scientific instruments in other culture areas is well developed. In the West, there are excellent museums of history of science and technology, such as the Museum of the History of Science at Oxford, the Adler Planetarium at Chicago, or the Astronomisch-Physikalische Kabinett at Kassel in Germany. Also the literature on medieval European instruments and instrument-makers is truly impressive, beginning with Ernst Zinner's classic study *Deutsche und Niederlaendische Astronomische Instrumente*.[15] Islamic astronomical instruments like the astrolabe and the celestial globe have been studied[16] or catalogued in several publications, notably in *A Computerized Checklist of Astrolabes* by Derek Price and his associates[17] and in the *Islamicate Celestial Globes, their History, Construction, and Use* by Emilie Savage-Smith.[18] These two surveys cover specimens manufactured in India as well. But the authors depended more on the scantily published notices rather than on physical examination in India, with the result that their listing of astrolabes and celestial globes in India is rather inadequate. Professor David King of the University of Frankfurt is currently engaged in cataloguing medieval European and Islamic astronomical instruments,[19] but his project will touch Indian instruments only in the periphery.

0.4 In India, no comparable work has been done on the existing instruments. While studying Jai Singh's masonry instruments in

15. Munich 1979 (reprint of the second revised edition).
16. Cf. in particular, King, David A., *Islamic Astronomical Instruments,* London 1987, and the bibliography therein.
17. Gibbs, Sharon L., Henderson, Janice A. & Price, Derek de Solla, *A Computerized Checklist of Astrolabes,* Yale University, New Haven, 1973.
18. Washington, D.C., 1985.
19. King, David A., Medieval Astronomical Instruments: A Catalogue in Preparation, *Bulletin of the Scientific Instrument Society*, 31, 3-7, 1991.

the early part of this century, Garrett and Kaye paid some attention to the portable instruments at Jaipur.[20] In 1921 Kaye also studied the four instruments acquired by the Archaeological Museum at Delhi.[21] The first one to attempt a survey all over India was Syed Sulaiman Nadvi. In his article of 1935, he threw light on a hitherto unknown family of astrolabe-makers of Lahore who produced astrolabes and celestial globes for four generations, and received patronage from the Mughal rulers from Humāyūn to Aurangzeb. Nadvi counted eight astrolabes and four celestial globes manufactured by the various members of this family between the years AD 1567 and 1663.[22] In the '70s, Professor S.N. Sen undertook a more systematic survey of "Astronomical Instruments of Historical Importance", and published a valuable study on the astrolabe.[23] Beyond this, nothing more came out of the project.[24] Moreover, it should be emphasized that nearly every study up till now concerned itself only with the astrolabes and celestial globes

20. Garrett, A. ff. & Guleri, Chandradhar, *The Jaipur Observatory and its Builder*, Allahabad 1902, pp. 60-65; Kaye, G.R., *The Astronomical Observatories of Jai Singh,* Calcutta 1918, pp. 16-36.
21. Kaye, G.R., *Astronomical Instruments in the Delhi Museum*, Calcutta, 1921.
22. Nadvi, Syed Sulaiman, Some Indian Astrolabe-Makers, *Islamic Culture*, 9, 621-631, 1935. Today, however, nearly ninety astrolabes and twenty-five celestial globes produced in this family are known, thanks to the efforts mainly of Professor Derek Price and Dr Emilie Savage-Smith.
23. Sen, S.N., The Astrolabe—A Case for Transmission of Technique of an Astronomical Instrument in Medieval India, in: W.H. Abdi et al (ed), *Interaction between Indian and Central Asian Science and Technology in Medieval Times*, New Delhi, 1990, vol. I, pp. 111-131.
24. Vijai Govind, Sen's assistant in this project, published two more papers. Govind, Vijai, A Survey of Medieval Indian Astrolabes, *Bharatiya Vidya,* 39.1, 1-30, 1979; Behari, Kailash & Govind, Vijai, A Survey of Historical Astrolabes of Delhi, *Indian Journal of History of Science*, 15, 94-104, 1980. In these two articles, there are serious flaws in the identification of instruments. For the sake of comprehensiveness, two other papers may be mentioned here. Dube, Padmakara, Astrolabes in the State Library of Rampur, *The Journal of the United Provinces Historical Society*, 4.1, 1-11, Pls. I-VII (October 1928); Stone, A.P., Astronomical Instruments at Calcutta, Delhi and Jaipur, *Archives internationales d'histoire des sciences,* 11, 159-162, 1958.

and ignored all other instruments. No doubt, these two are highly complex instruments and sought after by collectors, but for a history of Indian astronomical instruments, even simpler instruments deserve to be noticed.

0.5 Therefore, it is essential to make a comprehensive survey of all pre-modern Indian instruments preserved in private and public collections in India and abroad, and to prepare a descriptive catalogue with facsimile illustrations of all significant objects. Such a catalogue will serve as a useful research tool. It will supplement the data from the Sanskrit texts and may also add new variations not known to the texts. It may throw light on the geographical spread of instrument types and the methods and centres of their manufacture. For example, in the stores of the Jaipur Observatory, there are several astrolabes in various stages of completion, which yield interesting clues of the process of astrolabe making. Finally, it is hoped that the survey will lead to a better preservation of the instruments, especially in museums in India.

0.6 With these aims in mind, I made an exploratory survey, in 1991, of the instrument collections at Salar Jung Museum and State Museum of Archaeology, Hyderabad; Jantar Mantar Observatory, Hawa Mahal, City Palace Museum, and the Museum of Indology at Jaipur; Bharat Kala Bhavan and Sampurnanand Sanskrit University at Varanasi; and the Khuda Bakhsh Oriental Public Library at Patna. The results were highly encouraging. Everywhere I found many more items than I had anticipated. Equally gratifying has been the cooperation readily offered by the authorities of all these institutions.

Then in 1992-93 I had the opportunity to study and catalogue all the known Indian instruments in the US and UK. In the US, the main collections of Indian instruments are at Harvard University, Cambridge; Columbia University, New York; Adler Planetarium; Chicago; Time Museum, Rockford; and the Smithsonian, Washington.[25] The holdings of the Time Museum[26] and the

25. In these collections, there are over fifty Indian instruments, a report of which will be published elsewhere.

26. Turner, A.J., *Astrolabes, Astrolabe-related Instruments* (*The Time Museum: Catalogue of the Collection,* vol. I, part 1), Rockford 1985; idem, *Water-Clocks, Sand-Glasses, Fire-Clocks* (*The Time Museum: Catalogue of the Collection*, vol. I, part 3), Rockford 1984.

Smithsonian[27] are described in published catalogues as also the holdings of an important private collection owned by Leonard Linton.[28]

Indian instrument collections in Britain[29] are naturally the largest and well preserved.[30] In London, the British Museum, Victoria & Albert Museum, Science Museum, The Museum of the Worshipful Company of Clockmakers, and the National Maritime Museum at Greenwich possess many Indian instruments. There are also good collections at Oxford, in the Museum of the History of Science and in the Pitt Rivers Museum. Small collections exist at the Whipple Museum of History of Science, Cambridge; Welsh Museum, Cardiff; and the Royal Scottish Museum, Edinburgh.[31] The Museum of the History of Science at Oxford has also extensive archives containing photographs and descriptions of scientific instruments that came for sale in the various auction houses of London such as Christie's and Sotheby's.[32]

0.7 Now remains the daunting task of searching the length and breadth of India, locating private or public collections that may contain astronomical and time-measuring instruments, seeking the permission of the authorities to study the instruments and so on, which I hope to accomplish in the coming years. While the major Indian museums possessing instruments are known, information is

27. Gibbs, Sharon & Saliba, George, *Planispheric Astrolabes from the National Museum of American History,* Washington, D.C. 1984.
28. *Collection Leonard Linton et de divers amateurs, Auction Catalogue,* Paris 1980.
29. There are about one hundred items, the report of which will appear elsewhere.
30. Some of these were displayed in the Festival of India at London in 1972. See the Exhibition Catalogue compiled by R.G.W. Anderson, *Science in India: A 'Festival of India' Exhibition at the Science Museum, London, 24 March – 1 August 1982*, London 1982.
31. The task of locating historical scientific instruments is much facilitated by the recent publication of Holbrook, Mary, Anderson, R.G.W. & Bryden, D.J., *Science Preserved: A Directory of Scientific Instruments in Collections in the United Kingdom and Eire*, London 1992.
32. Thus I have seen a large part of Indian instruments abroad. There are still about thirty or more instruments in France and Germany, besides single specimens in Russia and the Middle-Eastern countries.

welcome about local museums where instruments may be lying unidentified or uncatalogued.

0.8 The proposed catalogue will contain a comprehensive account of all extant Indian instruments in all private and public collections in India and abroad, with historical surveys of the development of each instrument-type, its use and geographic spread, and a full technical description of each specimen with art-historical notes on the decorations and ornamentation. Since this will be the first work of its kind on Indian instruments, attempts will be made to provide also a full photographic documentation of each specimen.

0.9 For the sake of collecting material, I have tentatively classified the Indian instruments into the following categories:

1.1 Water Clocks, Outflow Type
1.2 Water Clocks, Sinking Bowl Type
2.1 Sun Dials with Vertical Gnomon
2.2 Sun Dials with Horizontal Gnomon
2.3 Equinoctial Sun Dials
2.4 Column Sun Dials
2.5 Other Kinds of Sun Dials
3.1 Sand Clocks, calibrated for *Ghaṭīs*
3.2 Sand-Clocks, calibrated for Hours
4.1 Quadrants, Sanskrit
4.2 Quadrants, Indo-Persian
5.1 Armillary Spheres, Sanskrit
5.2 Armillary Spheres, Indo-Persian
6.1 Ring Dials
6.2 Universal Ring Dials
7.0 Nocturnal-cum-Quadrant (*Dhruvabhrama-yantra*)
8.0 *Phalaka-yantras*, as described by Bhāskara II
9.1 Astrolabes, Indo-Persian, by Allāhdād Family
9.2 Astrolabes, Indo-Persian, by others
9.3 Astrolabes, Sanskrit, for several Latitudes
9.4 Astrolabes, Sanskrit, for a single Latitude
10.1 Celestial Globes, Indo-Persian, by Allāhdād Family
10.2 Celestial Globes, Indo-Persian, by others
10.3 Celestial Globes, Sanskrit
11.0 Other Instruments, inscribed in Sanskrit
12.0 Other Instruments, inscribed in Persian

0.10 In the following pages I shall present brief historical surveys of some major instrument types and invite attention to some important specimens I have come across.

1.1 I may begin with the common time-measuring devices. The oldest known instruments are the water clock (*Nālikā-yantra*) and the gnomon (*śaṅku*), both derived from Mesopotamian models. It is to these devices that the *Kauṭilīya Arthaśāstra* refers when it states that the king should divide the day and the night into eight periods each and perform various tasks in each period. The king is to measure these time divisions either with the water clock[33] or with the shadow of the gnomon.[34]

The water clock mentioned here is of the outflow type and this is the earliest type used in India:[35] it consisted of a hollow cylinder with a minute perforation at the bottom, through which it discharged a certain amount of water in course of 24 minutes or one-sixtieth part of a nychthemeron (*ahorātra*). It is this type of water clock which is known to the *Vedāṅgajyotiṣa*. Hollow tube is called *nala* in Sanskrit. The dimunitives *nālī/nāḍī; nālikā/nāḍikā* designate both the instrument and the time duration it measures.

1.2 Sometime about the fourth century AD, this type of water clock was replaced by another variety known as the sinking bowl type.[36] This consists of a hemispherical copper bowl with an extremely small perforation at the bottom. When this bowl is placed on the surface of water in a larger basin, the water enters the bowl from below through the perforation (see Fig. 1.1). As soon as the bowl is full, it sinks to the bottom of the basin with a clearly audible thud. The weight of the bowl and the size of the perforation are so adjusted that the bowl sinks sixty times in a day-and-night. That is

33. On the history of water clocks in India, see my forthcoming monograph, *Water Clocks and Time Measurement in India.*
34. *The Kauṭilīya Arthaśāstra*, ed & tr, R.P. Kangle, part 1, second edn, Bombay 1969, 1.19.6: *nālikābhir ahar aṣṭadhā rātriṃ ca vibhajec chāyāpramāṇena vā.*
35. On the classification of water clocks, see Needham, Joseph, *Science and Civilisation in China*, vol. 3: Mathematics and the Sciences of the Heavens and the Earth, Cambridge 1959, p. 315; Turner, A.J., *Water-clocks, Sand-glasses, Fire-Clocks* (see n. 26 above), p. 1.
36. Sarma, Sreeramula Rajeswara, Astronomical Instruments in Mughal Miniatures (see n. 12 above), pp. 241-243; see in this Volume, pp. 84-86.

Fig. 1.1. Water clock made of coconut shell, Government Museum, Madras. Photo: Courtesy of Professor S. Ramaratnam who floated it in a bucket of water and found that it takes exactly 24 minutes to sink.

to say, the duration of each immersion is 24 minutes. In Sanskrit, the bowl is called *ghaṭī* or *ghaṭikā* and these two terms designate also the duration of time measured by this device. The whole apparatus was accordingly called *Ghaṭī-yantra* or *Ghātikā-yantra.* It was described for the first time by Āryabhaṭa in his *Āryabhaṭasiddhānta.* After him, nearly every author of a *siddhānta* described it in his chapter on instruments.[37]

From the fifth century onwards, institutions for time-keeping are attested at Buddhist monasteries, royal palaces, town squares, etc., where time was measured constantly with this water clock and the passage of each *ghaṭī* and completion of each quarter of the day (*prahara*) or of the night (*yāma*) was broadcast regularly by means of drums and conch-shells. In the medieval period, the drum and

37. Several of these descriptions are collected in Subbarayappa, B.V. & Sarma, K.V., *Indian Astronomy: A Source-Book*, Bombay 1985, pp. 86-99.

the conch-shell were replaced by the gong, which was called *ghaḍiyāla* in the Middle Indic.

The institution was so popular that Muslim rulers adopted not only the device but also the time units *prahara, ghaṭī,* and its sub-multiple *pala*. Thus in the latter half of the fourteenth century, Fīrūz Shāh Tughluq, whose interest in sciences and engineering is well known, set up at his palace gate the *ṭās-i ghariyāla* (literally, "the bowl and the gong") which, in the words of his chronicler Shams-i-Sirāj 'Afīf, "aroused the wonder of people from Khurasan to Bengal." Fīrūz also had the picture of this water clock engraved on his coins and used them as ceremonial gifts to dignitaries.[38]

Two centuries later, the Mughal emperor Bābur gives a detailed description of this device in his memoirs. In every major town of Hindusthan, says he, there existed a class of people known as *ghariyālīs* whose profession it was to keep time with this water clock and announce it by means of the gong. Bābur not only adopted this device, he also made improvements in the mode of announcing the *pahars* and *gharis*.[39]

Abū 'l-Fazl's *Ā'īn-i Akbarī* has a long account of the Institution of Ghariyālā. It informs that time keeping and announcing was the royal prerogative under Akbar.[40] More important, Akbar's painters depicted the water clock in two miniature paintings. Indeed these are the only pictorial representations of this device.[41]

Subsequently, minor nobles also began to maintain water clocks at their palace gates and to regularly announce time. This continued until the beginning of this century. Even now it is occasionally used for ritual purposes. In 1979, it is reported that in the mausoleum of Shaikh 'Usmān, popularly known as Lāl Shahbāz Qalandar, at the city of Sehwan in Sind, time is still measured by the water clock

38. Cf. Sarma, S.R. & Alam, Ishrat, Announcing Time: the Unique Method at Hayatnagar, 1676, *Proceedings of the Indian History Congress, 52nd Session, New Delhi, 1991-92,* Delhi 1992, pp. 426-431; reprinted in this Volume, pp. 136-42.
39. Beveridge, Annette Susannah (tr), *Bābur-Nāma (Memoirs of Bābur)*, Delhi 1979, vol. I, pp. 516-517.
40. Jarret, H.S. (tr), *Ā'īn-i Akbarī,* vol. III, Delhi 1977, pp. 17-18.
41. Cf. Sarma, Sreeramula Rajeswara, Astronomical Instruments in Mughal Miniatures (n. 12 above), p. 242; Pls. 2-3; see in this Volume, pp. 111, 112; Figs. 4.2-3.

for performing the various rituals. Possibly the practice is still continuing.[42]

1.3 Considering the universal popularity of this water clock, there ought to be hundreds of specimens extant. But in India brass or copper objects which are no longer in use are immediately recycled. Consequently not many seem to have survived. The largest collection is in the Pitt Rivers Museum at Oxford. In India, I have counted so far only six.

Except in one case, it has not been possible for me to actually measure time by floating the bowls on water. However, most of the bowls I have seen appear to have been made for measuring one *ghaṭī*. But how did one measure fractions of a *ghaṭī?* It is of course possible to divide the inner surface of the bowl empirically, if not into 60 divisions, at least into 10 or 6 units. In recent literature, there are references to such graduated bowls[43] but I have not seen one so far. Likewise, not a single basin (Sanskrit *kuṇḍa*) in which the perforated bowl is made to float seems to have been preserved. Nor unfortunately were any gongs collected.[44] In literature larger cups for measuring more than one *ghaṭī* are attested, but none has been noticed so far.

2.1 I have mentioned that the other common time-measuring device was the gnomon or sundial. In the course of centuries, several kinds of sundials developed with a variety of dials and

42. Baloch, N.A., Measurement of Space and Time in the lower Indus Valley of Sind, in: Said, Hakim Mohammed (ed), *History and Philosophy of Science: Proceedings of the International Congress of History and Philosophy of Science, Islamabad, 8-13 December, 1979,* Karachi n.d., part I, pp. 168-196, esp. 186-194.
43. Gilchrist, John, Account of the Hindustanee Horometry, *Asiatick Researches*, 5, 81-89, 1795; Ali, Mrs. Meer Hassan, *Observations on the Mussulmans of India*, (first published: London 1832) Karachi 1974, p. 55; Thurston, Edgar, *Ethnographic Notes on Southern India*, 1907, reprint: Delhi, 1975, p. 565.
44. In the last century, a German nobleman collected a gong, but its current whereabouts are unknown. Cf. Schlagintweit-Sakünlünski, Hermann von, Eine Wasseruhr und eine metallene Klangscheibe alter indischer Construction, *Sitzungsberichte der mathematisch-physikalischen Classe der k.b. Akademie der Wissenschaften zu Muenchen*, 1, 128-139, 1871.

Fig. 1.2. Vertical gnomon on a horizontal dial, dated 1294 AH/AD 1875, Firdaus Manzil, Mangal Talab, Patna.

gnomons. The most common form is the horizontal dial with a vertical gnomon, which can also be made temporarily by inserting a stick into the ground. So far I have seen permanent structures of this kind only in mosques (Fig. 1.2). In fact, every mosque is supposed to have a sundial to determine the times of prayer. Here too there can be a wide variety in the design. But these are being removed at an alarming rate. There is an urgent need of preserving these scientific instruments.

2.2 In late medieval times, a variant of the horizontal dial with a triangular gnomon seems to have become popular. The hypotenuse of the gnomon is elevated from its base by an angle equal to the local latitude, and thus it lies in the plane of the celestial equator. The *Yantraprakāra of Sawai Jai Singh* calls it *Palabhā-yantra*, and teaches the method of constructing this type of sundial for the latitude of Delhi (28;39°) and for that of Amber (27°).[45] A small *Palabhā-yantra* is erected on the top of the *Nāḍīvalaya-yantra* of

45. *Yantraprakāra of Sawai Jai Singh* (n. 3 above), pp. 26, 76-77.

the Jaipur Observatory. I have also seen a very beautiful specimen in the Khanqah Emadia at Patna. It was designed by Janab Faridul Emadi, the present head of the Khanqa.[46]

3.1 Horizontal gnomon is usually attached to vertical walls. But in one variant, it is inserted into a hole at the top of a circular or prismatic column of wood. Below the hole is a scale to measure the shadow of the gnomon in terms of hours or *ghaṭīs* from the sunrise or up to the sunset. But as the sun's altitude varies according to seasons, separate scales with separate holes are provided for each season. Usually there is one scale for each month or each pair of months. The instrument is said to have been invented in the Islamic World whence it spread both to Europe and India. In India it was introduced under the name *Cābuk* or whip instrument. Sanskrit texts literally translate the word as *Kaśā-yantra* or *Pratoda-yantra*, or just transliterate as *Cābuka-yantra.* The first description in India occurs in Rāmacandra's *Yantraprakāśa* of 1428.[47]

In India, I have seen so far only one crude specimen in the stores of Jaipur Observatory. It is about three feet long and has a scale for each month. The scales are painted on the wood but now the paint has all peeled off.[48] In the US, the Time Museum at Rockford has a highly ornate column dial made of ivory. In the Museum of the History of Science at Oxford, there is an exquisitely crafted column dial made of damascened steel. All these three were manufactured in western India.

3.2 A somewhat different variety of wooden column dial was very popular in Nepal-Darjeeling region in the last century.[49] I have seen some ten specimens in the various museums in UK. Here the numbers on the scales and the names of the months for which each scale is meant are carved in relief in the wood (Fig. 1.3). Apparently these dials were called Āṣāḍha sticks, because the longest day

46. There is a small marble *Palabhā-yantra* in the City Palace Museum, Jaipur, and another at Bharat Kala Bhavan, Varanasi.
47. Cf. Sarma, Sreeramula Rajeswara, Rāmacandra Vājapeyin's Contributions to Astronomy and Mathematics (n. 9 above); see also Sharma, S.D., *Pratoda Yantra by Ganesha Daivajna*, Kurali 1982.
48. Cf. Sarma, Sreeramula Rajeswara, Portable Instruments at the Jaipur Observatory (n. 4 above).
49. Cf. Winter, H.J.J., A Shepherd's Time-Stick, Nāgarī Inscribed, *PHYSIS*, 4, 377-384, 1964.

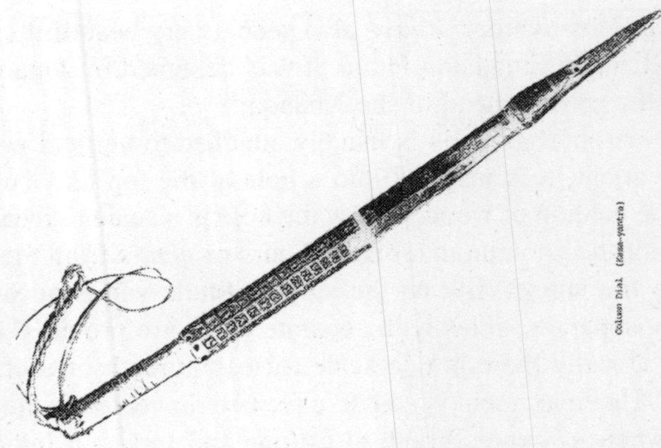

Fig. 1.3. Carved wooden column dial, a composite picture.

occurs in the month of Āṣāḍha and consequently the scale for this month is the longest. Only one of these, at the Horniman Museum, Forest Hill, London, is signed. It carries an inscription to the effect that Jemangala engraved it on Saṃvat 1941 Pauṣa sudi 6.

4.0 Coming to the astronomical instruments proper, one instrument with a long history is the quadrant (*Turīya-yantra*) (Fig. 1.4). Brahmagupta includes it in his list though he prefers a

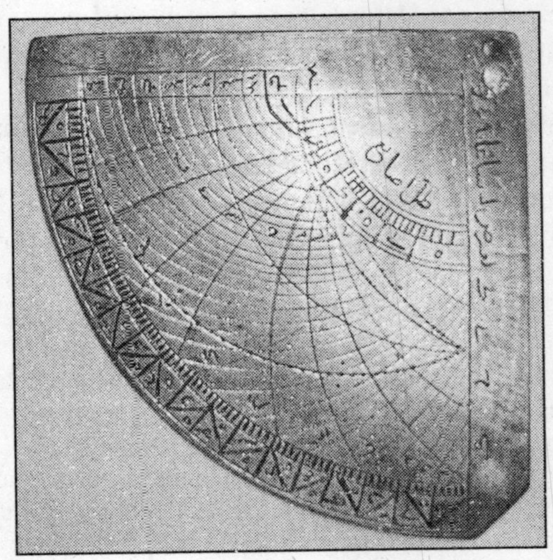

Fig. 1.4. Indo-Persian quadrant, unsigned, undated.
National Maritime Museum, Greenwich, Acc. No. 9204.

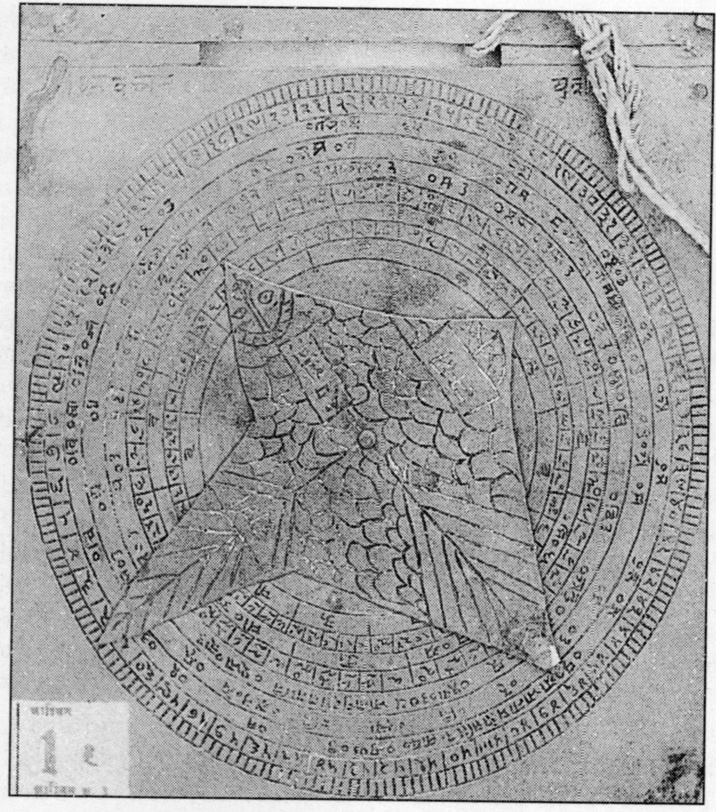

Fig. 1.5. *Dhruvabhrama-yantra*, unsigned, undated.
Jai Singh's Observatory, Jaipur.

double-quadrant for measuring the altitude and time. There are not many specimens extant, as this instrument is incorporated on the one hand in the *Dhruvabhrama-yantra* (see 5.0) and on the other in the astrolabe. Outside India, there are two or three Sanskrit quadrants made in the nineteenth century. In India I have seen so far one at Jaipur and another at Varanasi. A Turkish style wooden quadrant made for the latitude of Samarqand is in Bombay.[50]

5.0 The most ingenious instrument of pre-Islamic India is the *Dhruvabhrama-yantra* which functions as a kind of nocturnal (Fig. 1.5). It was described for the first time by Padmanābha in AD

50. Khareghat, K.P., *Astrolabes*, Bombay 1950, pp. 75-82.

1423 and may have been invented by him. It consists of an oblong metal plate with a horizontal slit at the top. Loosely pivoted to the centre of the plate is a metal index with four indicators projecting into the four directions. A plumb is suspended from the southernmost end so that it always points downwards. Around the centre are concentric circles containing various scales. At night, the instrument is so held that one can see through the slit the pole star and β Ursae Minoris. When these two stars are sighted in a straight line by appropriately tilting the instrument, the eastern indicator will point to the *lagna* or ascendant for this moment, the northern indicator will show the culminating point and the western indicator the sidereal time in *ghaṭīs* and *palas*. Thus with this instrument, one can read off from the dial, for any given moment at night, the corresponding time, ascendant, and culmination.[51]

The reverse side of this instrument usually contains a horary quadrant (*Turīya-yantra*) for measuring time during the day. In a specimen at Jaipur, there is a scale of degrees in the outermost arc, on which the solar altitude can be measured. In the other arcs there are separate *ghaṭī*-scales for directly measuring time in different seasons. I have seen a few beautiful specimens of this instrument in UK and USA. In India, there is one at Jaipur and another at Varanasi. The Khuda Bakhsh Oriental Public Library at Patna has, what may be called an Islamic version, one side of which is called *Rūznumah* for the use in the daytime and the reverse is calibrated as *Shabnumah* for the night time. The piece was made in 1804 at Bareilly by Naṣīr al-Dīn Hasan.

6.0 The most important medieval scientific instrument, however, is the astrolabe.[52] The astrolabe and the celestial globe were invented in Hellenistic antiquity but they reached their perfection in the Islamic World, whence they were transmitted westwards to Europe and eastwards to India. In the Islamic World these two were

51. Cf. Garrett, op. cit., pp. 62-64; Pl. X.
52. The literature on the astrolabe is very extensive. The best accounts are Hartner, Willy, The Principle and Use of the Astrolabe, in: idem, *Oriens-Occidens*, Hildesheim 1968, pp. 287-311; North, J.D., The Astrolabe, *Scientific American*, 230, 96-106, January 1974. On the history of the astrolabe in India, see Sarma, Sreeramula Rajeswara, Astronomical Instruments in Mughal Miniatures (n. 12 above), pp. 237-241; see in this Volume, pp. 79-84.

the most popular instruments, and the usual mode of describing a good astronomer was to say that he was adept in the use of the astrolabe and the celestial globe.

While the celestial globe was a teaching instrument, the astrolabe is a highly versatile observational and computational instrument. With it time can be determined both in the daytime and at night, both in seasonal hours and in equal hours. One can measure the altitudes of heavenly bodies or the heights of distant objects. More important still is that it works like an analogue computer. For any given moment, one can read off directly from the dial the four points of the ecliptic such as the ascendant, etc., which are important in horoscopy. Jābir al-Ṣūfī is said to have enumerated a thousand problems in spherical trigonometry that can be solved by means of the astrolabe.

6.1 When was the astrolabe introduced into India? Al-Bīrūnī, who wrote a number of tracts on this instrument, must have brought it to India in the eleventh century, and explained its working principles to his Hindu interlocutors at Multan. In the next centuries, Muslim scholars migrating from Central Asia to the court of the Sultans of Delhi brought astrolabes with them.

By the mid-fourteenth century, the instrument was sufficiently well known to be mentioned in a Persian work of fiction called *Basāṭīn al-Uns* written at Delhi.[53] In the same century, astrolabe manufacture also appears to have commenced in India under the active patronage of Fīrūz Shāh Tughluq. The *Sīrat-i Fīrūz Shāhī*, an anonymous chronicle of his rule composed in 1370, has a long account on the astrolabes manufactured under instructions from Fīrūz.[54] However, none of these survive today.

6.2 The earliest surviving astrolabes pertain to the Mughal period. Of the Mughal emperors, Humāyūn is said to have "extraordinary excellence in the astrolabe, globe and other instruments of the observatory". Under his patronage, manufacture of astrolabes and celestial globes commenced at Lahore. One family, in particular,

53. Siddiqi, Iqtidar H., Basāṭīn al-Uns: A Source of Information on the Sultanate of Delhi under the early Tughluq Sultans, *Quarterly Journal of the Pakistan Historical Society*, 36.4, 293-302, 1988.

54. Cf. Sarma, Sreeramula Rajeswara, Sulṭān, Sūri and the Astrolabe, *Indian Journal of History of Science*, 35, 129-147; reprinted in this Volume, pp. 179-98.

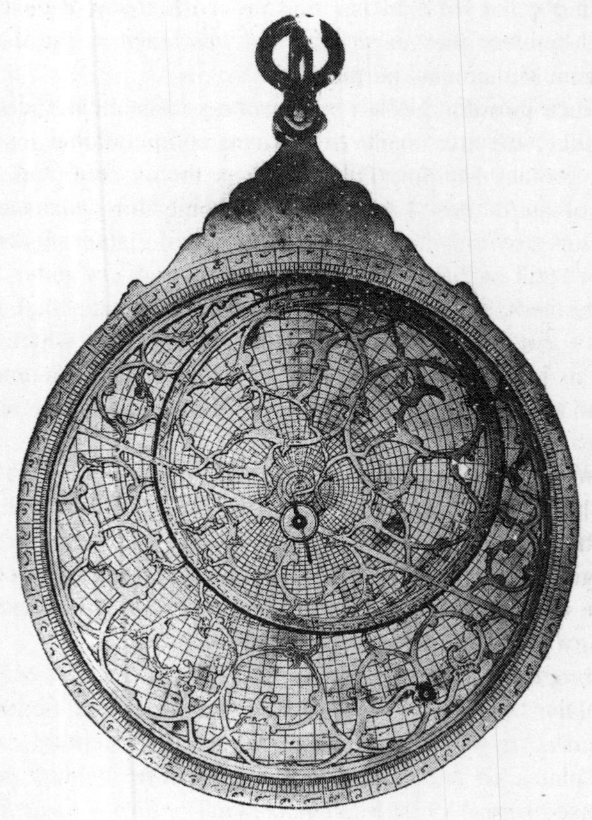

Fig. 1.6. Astrolabe by ʻĪsà ibn Allāhdād Asṭurlābī Humāyūnī Lāhūrī, 5 plates, not dated. Personal collection of Mr John N Didcock, Ely, Cambs, UK. Photo: Courtesy of Mr Didcock.

distinguished itself in this pursuit. Seven members of this family, belonging to four generations, signed their names on about 87 astrolabes and 21 celestial globes (Fig. 1.9). There are also many unsigned pieces which can be attributed to this family. The patriarch of this family is Ustād Shaikh Allāhdād Asṭurlābī Humāyūnī Lāhūrī. He may have been the royal astrolabe maker of Humāyūn and hence received the title Asṭurlābī Humāyūnī, although the only dated piece by him was made in 1567, i.e. eleven years after Humāyūn's death. It is now in the Salar Jung Museum at Hyderabad.

Allāhdād's son Mullā 'Īsá is known through five astrolabes produced between the years 1600 and 1604 (Fig. 1.6). Of his two sons, Qā'im Muḥammad perfected the art of casting celestial globes in one piece through the lost wax process. Four of his globes are extant today. He also made five astrolabes, two of which jointly with his brother Muqīm. The latter has to his credit the largest number of astrolabes. I have counted some thirty-six astrolabes manufactured by him during the half a century of creativity between the years 1609 and 1659.

6.3 The most prolific and versatile member of this family is Qā'im's son Ḍiyā' al-Dīn Muḥammad, whose period of creative activity lasted from 1637 to 1680, and overlapped the reigns of Shāh Jahān and Aurangzeb. His oeuvre consists of about 29 astrolabes and 15 celestial globes—perhaps the largest number ever signed by a single instrument-maker of pre-industrial times. Today, these pieces are dispersed in all parts of the world: India, Egypt, Russia, France, Germany, UK and USA. His prolific work is distinguished by superb craftsmanship and innovation in design. Here one can see the harmonious combination of art, science and technology at its perfection.

6.4 Chronologically the last and technically the most remarkable of Ḍiyā' al-Dīn's astrolabes is the universal astrolabe he manufactured in 1680 for Nawāb Iftikār Khān of Jaunpur. Here separate plates are not needed for different latitudes, and the same plate is used for all localities. The original prototype was designed by Ibn al-Zarqāllu of Cordova in the eleventh century. There are not many specimens of this type extant today. Ḍiyā' al-Dīn's creation is nearly the largest and most elaborate one of all the extant specimens. It measures 555 mm across and is one of its kind in the entire east.[55]

While the piece itself is thus unique, its subsequent history is also of great interest for the transmission of ideas. Within half a century of its manufacture, it was acquired by Sawai Jai Singh, who used it for the instruction of the Hindu astronomers at his court. On the crown of the instrument, its name was engraved in Sanskrit/Rajasthani: *Yantra Jarakālī Sarvadeśī.* A copper plaque

55. Kaye, G.R., *The Astronomical Observatories of Jai Singh*, pp. 27-30.

was attached to it with all its functions engraved in Rajasthani. Sanskrit equivalents of a number o: star names were also engraved on it. More important still is that, inspired by this piece, Sawai Jai Singh caused a Sanskrit manual to be composed on the construction and use of this universal astrolabe.[56]

6.5 Thus the creations by the various members of the Allāhdād family are reasonably well covered. Besides this family, there were also other instrument-makers who produced Indo-Persian astrolabes and celestial globes. As the survey progresses, it is hoped, that more names will come to light. The tradition continued until the middle of the nineteenth century. The last great practitioner was Lālah Bahlūmal Lāhūrī, whose work will be discussed presently.

7.0 The astrolabe was received by Hindus and Jainas with great enthusiasm. I have mentioned that Mahendra Sūri wrote the first Sanskrit manual in 1370 under the auspices of Fīrūz Shāh Tughluq at Delhi. Mahendra Sūri was so impressed by the versatile functions of the astrolabe that he called it *Yantrarāja*, "king of astronomical instruments", and it is under this name that the astrolabe came to be known in Sanskrit and other Indian languages. Hindu astronomers who came later were no less enthusiastic. Between the fourteenth and eighteenth centuries, more than a dozen manuals were composed in Sanskrit on the astrolabe. These include the Sanskrit rendering of Naṣīr al-Dīn al-Ṭūsī's Persian manual in twenty chapters, popularly known as the *Bist Bāb*.[57]

7.1 The composition of these Sanskrit manuals must have been accompanied by the manufacture of Sanskrit astrolabes as well, that is, astrolabes with legends in Sanskrit and the time scale divided into *ghaṭīs* instead of hours (Fig. 1.7). The earliest known Sanskrit astrolabe was manufactured in 1607 by one Dāmodara. It came for sale at Sotheby's in 1976, but its present whereabouts are not known. Consequently, the earliest extant astrolabe in a public collection today is the one manufactured by Maṇirāma in 1643. It is

56. Cf. Sarma, Sreeramula Rajeswara, The Ṣafīḥa Zarqāliyya in India, in: Josep Casulleras & Julio Samsó (ed), *From Baghdad to Barcelona: Studies in Islamic Exact Sciences in Honour of Juan Vernet*, Barcelona 1996, pp. 719-735; reprinted in this Volume, pp. 223-39.

57. *Yantrarājavicāra-viṃśādhyāyī of Nayanasukhopādhyāya*, ed. Bibutibhushan Bhattacharya, Varanasi 1979.

Fig. 1.7. Sanskrit astrolabe, single plate for the latitude 22;35,39° N, dated 1941 VS/AD 1884, unsigned. Formerly in the collection of Saul Moskowitz, Marblehead, Mass., USA. Photo: Courtesy of Professor David Pingree.

now at the Royal Scottish Museum of Edinburgh. Pitt Rivers Museum of Oxford has one astrolabe made in 1673 for the astrologer Indrajī.

7.2 The manufacture of Sanskrit astrolabes received a great impetus from Sawai Jai Singh who had himself authored a book on its construction.[58] He had apparently a *kārkhānā* at Jaipur for the

58. *Yantrarājaracanā of Jayasiṃha*, ed. Kedāranātha Jyotirvid, Jaipur 1953.

manufacture of astrolabes.[59] This manufactory produced some highly ornate astrolabes with multiple plates. But its main product were astrolabes with a single plate calibrated for 27° N, which is the latitude of Jaipur. Hindu astronomers of Rajasthan found the astrolabe an excellent teaching instrument. Therefore, astrolabes continued to be manufactured in the 18th and 19th centuries in Rajasthan. These are mainly single plate astrolabes, calibrated for a single latitude, usually 27°. Besides Jaipur, Kuchaman (27;10° N; 74;54° E) seems to have been an important centre of production of these astrolabes.[60]

8.1 The celestial globe, the other important Islamic instrument, is a late comer to India.[61] It was probably introduced by Humāyūn into India (Fig. 1.8). As mentioned already, the descendants of Allāhdād manufactured globes of good workmanship. Allāhdād's grandson, Qā'im Muḥammad perfected the art of casting globes in one piece by the cire perdue process, and his son Ḍiyā' al-Dīn excelled in this technique. Besides some 15 ordinary celestial globes, he also designed a very unusual one for Aurangzeb in 1679. The surface of the globe was cut *à jour*, like the spider of the astrolabe, leaving out the outlines of the constellation figures, great circles, tropics and arctic circles. Star positions are indicated by small perforations. When lit from inside, the globe would present an illuminated view of the celestial sphere.

59. Cf. Sarma, Sreeramula Rajeswara, Portable Instruments at the Jaipur Observatory (n. 4 above).
60. In the archives of the Museum of the History of Science at Oxford, there is a large collection of photographs of Sanskrit astrolabes that came for sale in the auction houses of London in the past two or three decades. Though the current location of these astrolabes is not known, from the photographs I could build up a large repertoire of the names of persons who designed, manufactured or owned Sanskrit astrolabes.
61. Cf. Savage-Smith, Emilie, *Islamicate Celestial Globes: Their History, Construction, and Use*, Washington, D.C. 1985; Sarma, S.R., Ansari, S.M.R. & Kulkarni, A.G., Two Mughal Celestial Globes, *Indian Journal of History of Science*, 28, 55-65, 1993; reprinted in this Volume, pp. 294-307; Sarma, Sreeramula Rajeswara, From al-Kura to Bhagola: On the Dissemination of the Celestial Globe in India, *Studies in History of Medicine and Science* 13.1, 69-85, 1994; reprinted in this Volume, pp. 275-93.

Fig. 1.8. Celestial globe by Bahlūmal Lāhūrī, dated 1258 AH/AD 1842. The Science Museum, London, Acc. No. 1985-1257.

8.2 Celestial globes were manufactured outside this family also. In fact, the earliest extant celestial globe was produced by one 'Alī Kashmīrī ibn Lūqmān during the reign of Akbar in 998 AH/AD 1589. Muḥammad Ṣālih of Thatta, a contemporary of Ḍiyā' al-Dīn, is known to have produced at least three globes.[62] Like the astrolabes, celestial globes also continued to be produced until the middle of the last century.

8.3 To judge from the surviving examples, the celestial globe was not popular among the Hindus as the astrolabe had been. So far, I have come across only three specimens, two at Jaipur and one at New York.

62. Cf. Sarma, S.R. et al, Two Mughal Celestial Globes (n. 61 above).

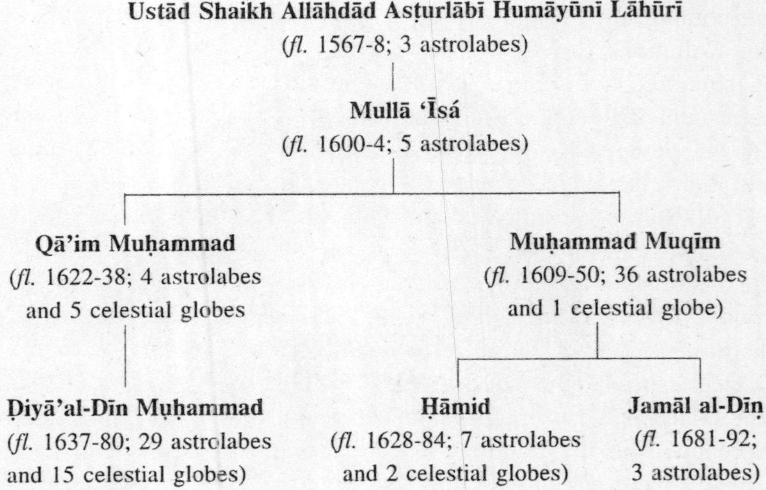

Fig. 1.9. The astrolabist family of Lahore and their instruments.

9.0 I have said that the survey may throw up new variants not mentioned in the texts, and it has indeed done so. Likewise more and more details are emerging about hitherto little known instrument-makers. One such person is Lālah Buhlovarmā or Buhlomalla or Bahlūmal Lāhūrī, who flourished in the first half of the nineteenth century. A heir to the metallurgical techniques of the Allāhdād family and, at the same time, well-versed in the Sanskrit tradition of instrumentation, he produced astrolabes and celestial globes with legends either in Arabo-Persian or in Sanskrit. He also crafted *Dhruvabhrama-yantras* and *Turīya-yantras*, signing his name in pretty Sanskrit verses. In Bahlūmal then we can see the true and perhaps the last representative of both the Islamic and Hindu traditions of scientific instrumentation.

10.0 The survey also revealed that, although the telescope is widely available, instruments for naked-eye observation continued to be produced in the 19th century and some of these were copies of European prototypes, with legends in Sanskrit for the Hindu astrologers or in Persian for Muslim astrologers. At the Khuda Bakhsh Library of Patna, there is a universal ring dial engraved in Sanskrit, and an adjustable equinoctial dial, engraved in Persian. As I proceed with this survey, more and more pieces of this nature are coming up. I found two more Sanskrit universal ring dials at the Columbia University, New York, and at the Smithsonian, Washington, D.C. The Adler Planetarium at Chicago has a very large adjustable equinoctial dial. When more such specimens are known, perhaps it will be possible to identify the manufacturers and their clients and to trace the context of this transmission.

11.0 Before closing this presentation, I should like to invite attention to an intriguing feature about the geographical spread of the instrument types. The sinking bowl type of water clock, of course, was used throughout South Asia. But all other pre-Islamic instruments, leaving aside the wooden column dials from the Nepal-Darjeeling area, emanate from the Rajasthan-Gujarat region only.[63] The Islamic instruments like the astrolabe and the celestial globes were made in the Lahore-Delhi area, and do not seem to have spread to the east and south. The outermost limit in the east is Tikari (24;57° N; 85;53° E) in Bihar, where Ghulām Ḥussain Jaunpūrī made a celestial globe in 1816[64] (and perhaps also astrolabes). The southern limit is Aurangabad, where a Muhammad Mūsá Aṣturlābī flourished in the 18th century. A celestial globe made by his grandson Muḥammad Faḍlullāh in 1808 (presumably at Aurangabad) survives today. This is mounted on a European style stand into which a magnetic compass is incorporated.[65] On the other hand, 19th century Sanskrit/Persian copies of European instruments were produced at Delhi, Bareilly and Patna. Equally intriguing is the following: Indo-Persian astrolabes and celestial globes produced in the 16th, 17th and 19th centuries have been found but not a single piece from the 18th century. Perhaps we will

63. Kerala has a tradition of observational astronomy, and there is a specific mention of an armillary sphere erected at the observatory of Mahodayapura in the ninth century, Cf. Sarma, K.V., *Observational Astronomy in India*, Calicut 1990, pp. 41-42. Do any instruments survive in Kerala today?

64. Cf. Ansari, S.M.R. & Sarma, S.R., Ghulām Ḥusain Jaunpūrī's Encyclopedia of Mathematics and Astronomy, *Studies in History of Science & Medicine*, 14, 77-93, 1999-2000.

65. Preserved at the Salar Jung Museum, Hyderabad. Cf. Sreeramula Rajeswara Sarma, *Astronomical Instruments in the Salar Jung Museum*, Hyderabad 1996, pp. 27-28, Fig. 15.

be able to answer these questions when the survey is complete. In the meantime, the author will welcome suggestions and, in particular, information about less known instrument collections.

ACKNOWLEDGEMENTS

This project would not have progressed this far without the cooperation of innumerable persons who went out of their way to help me, in particular the authorities and staff of all museums and libraries that I had visited. While acknowledgements will be made to all of them individually in the final catalogue, I may make a special mention here of the following scholars for their advice, encouragement and generous help: Dr R.G.W. Anderson, London; Professor S.M.R. Ansari, Aligarh; Dr A.K. Bag, New Delhi; Professor Owen Gingerich, Cambridge, Mass.; Professor David A. King, Frankfurt; Mr F.R. Maddison, Oxford; Professor David Pingree, Providence; Professor G.L. 'E. Turner, Oxford.

2

Astronomical Instruments in Brahmagupta's Brāhmasphuṭasiddhānta

Brahmagupta (b. AD 598) is well known as an excellent astronomer and mathematician. His two works, the *Brāhmasphuṭasiddhānta* (*BSS*)[1] and the *Khaṇḍakhādyaka,* written respectively in AD 628 and 665, exercised a great deal of influence both within India and outside. It is through the Arabic translations or adaptations of these two texts that the Islamic World became acquainted with Indian astronomy.[2] He also occupies a prominent place in the history of observational astronomy, for the *BSS* is the first extant text to contain a systematic discussion on the construction and use of a large number of astronomical instruments. It is, of course, true that before him Āryabhaṭa I (b. AD 476) had described some instruments

First published in: *Indian Historical Review,* 13 (1986-87) 63-74.

I wish to gratefully acknowledge the financial assistance generously provided to me by the Indian Council of Historical Research, New Delhi, for attending the Seventh World Sanskrit Conference held at the Kern Institute, Leiden, in August 1987, where this paper was presented. It is worthwhile recalling here that Professor H. Kern, after whom the Institute is named, brought out the first edition of the *Āryabhaṭīya* (Leiden, 1874) and also the first edition (Calcutta, 1865) and English translation (*JRAS,* 1870-75) of Varāhamihira's *Bṛhatsaṃhitā.*

1. I use the *editio princeps* by Sudhakara Dvivedi (Benares, 1902). Dr A. K. Bag, Indian National Science Academy, New Delhi, has kindly sent me a photocopy of the twenty-second chapter. The second edition by Ram Swarup Sharma (New Delhi, 1966) in four heavy volumes is quite innocent of textual criticism and even of proofreading.
2. On the dissemination of Indian astronomy in the Islamic World, notably through Brahmagupta's works, see Fuat Sezgin, *Geschichte des Arabischen Schrifttums,* Band VI. Astronomie bis ca. 430 H. (Leiden, 1978), pp. 116-120.

in his *Āryabhaṭasiddhānta*. This text is no more extant, but the portion on instruments survives in quotations, notably in Rāmakṛṣṇa Ārādhya's commentary (AD 1372) on the *Sūryasiddhānta*.[3] Varāhamihira (*c.* AD 550) also briefly mentioned a few instruments in his *Pañcasiddhāntikā* (14.19-32). But the instruments of these two astronomers did not find acceptance in later times as those discussed by Brahmagupta did.

In the twenty-second chapter of the *BSS*, called appropriately *Yantrādhyāya*, Brahmagupta not only describes many instruments, but also teaches methods of computing various astronomical elements from the readings taken with these instruments. These instruments and computational techniques were adopted in almost all later *siddhāntas* like the *Śiṣyadhīvṛddhida* of Lalla (eighth century), the *Siddhāntaśekhara* (AD 1039) of Śrīpati and the *Siddhāntaśiromaṇi* (AD 1150) of Bhāskara II. Therefore, in order to understand the nature and function of astronomical instruments in pre-Islamic India, it is essential to study the *BSS*.[4]

The instruments discussed by Brahmagupta can be divided into four groups: (i) accessories, (ii) astronomical instruments proper for measuring time and observing celestial bodies, (iii) instruments that turn automatically for one day, and (iv) those that rotate perpetually.

3. Collected by K.S. Shukla in "Āryabhaṭa I's Astronomy with Midnight Day–reckoning", *Gaṇita*, vol. XVIII (1967), pp. 83-105; and "Glimpses from the *Āryabhaṭasiddhānta*", *Indian Journal of History of Science*, vol. XII (1977), pp. 181-86. The Sanskrit stanzas on the instruments together with Shukla's translation are reprinted in B.V. Subbarayappa & K.V. Sarma, *Indian Astronomy: A Source-Book* (henceforth *Source-Book*) (Bombay, 1985), pp. 86-99, 182.
4. Little attention has been paid to Indian astronomical instruments before Jai Singh. Cf. S.R. Das, "Astronomical Instruments of the Hindus", *IHQ*, vol. IV (1928), pp. 259-69; R.N. Rai, "Astronomical Instruments", *Indian Journal of History of Science*, vol. XX (1986), pp. 308-36; Sankara Balakrishna Dikshita, *Bhāratīya Jyotiṣa*, tr. into Hindi by Sivanatha Jharakhandi (Lucknow, 1963), pp. 452-68; D.M. Bose et al, *A Concise History of Science in India* (New Delhi, 1971), pp. 124-26; David Pingree, *Jyotiḥśāstra: Astral and Mathematical Literature* (Wiesbaden, 1981), pp. 52-54; *Source-Book*, pp. 74-99.

I

Under accessories (*saṃsādhana*), Brahmagupta enumerates eight items: water, a pair of compasses (*bhrama*), plumb-line (*avalamba*), hypotenuse (*karṇa*), shadow (*chāyā*), midday (*dinārdha*), the sun and the local latitude (*akṣa*).[5] Āryabhaṭa mentions only the first four,[6] to which Brahmagupta adds, apparently for the sake of comprehensiveness, the four natural phenomena, the presence or the knowledge of which is required while using the astronomical instruments. However, the first four deserve our attention as they refer to primitive geometrical tools or methods for drawing circles or for aligning an instrument or a figure in all the three dimensions, namely, horizontal, vertical and lateral. That is to say, whether a plane is horizontal or not is tested by means of water, and whether a plane or line is truly vertical or not is ascertained with the plumb-line. By "lateral" (*tiryak*) is meant the following: when a straight line is extended sideways to enclose an area in the form of a triangle or rectangle, a pair of hypotenuses are needed to complete the figure. Their measurements constitute the fourth accessory called *karṇa*. Brahmagupta does not describe any of these methods or tools. However, the commentaries on the *Āryabhaṭīya* offer some interesting information.

These commentaries do not seem to be familiar with any simple level instrument. Instead, they recommend two methods of using water for determining the horizontality. Bhāskara I, who completed his commentary on the *Āryabhaṭīya* in AD 629, states:

> When there is no wind, place a jar full of water upon a tripod on the ground which had been made plane by means of eye or thread, and bore a [fine] hole [at the bottom of the jar] so that water may have a continuous flow. Where the water falling on the ground spreads in a circle, there the ground is in perfect level; where the water accumulates after departing from the circle of water, there it is low; and where the water does not reach, there it is high.[7]

5. *BSS*, 22.6-7.
6. *Āryabhaṭīya, Gaṇita,* 13.
7. K.S. Shukla, ed, *Āryabhaṭīya*, with the Commentary of Bhāskara I and Someśvara (New Delhi, 1976), p. 87. The translation is from Ibid., p. 55.

Since this was written a year after Brahmagupta completed the *BSS* and since both are from the same geographical area—Brahmagupta belonged to Bhillamāla and Bhāskara to Valabhī—it is likely that Brahmagupta was familiar with this method. The second method is described by Parameśvara who lived in Kerala in the fifteenth century:

First make the ground level by sight or by using a rope. Then draw a circle on the ground. . . . [With the same centre] draw another circle around the first one at a distance of one or two inches from the first. Dig the ground between the two circumferences and thus make a channel. Fill the channel with water. If the water is up to the level of the ground all around, then the ground is plane; where the ground is low, the level of water will be higher, and so on.[8]

Again, we owe to Parameśvara the only description of preparing a pair of compasses that is available in Sanskrit. He recommends as follows:

Get hold of some straight stick. Tie its top firmly with a rope, then splice the stick from bottom to top so that there are now two sticks. Sharpen their [lower] ends. Thus is made a pair of compasses with their points downwards. Then insert [another] stick between the two sticks and thus make the mouth of the compass wide. By moving the inserted stick up and down, the mouth of the compass can be widened so as to equal the radius of the desired circle. Then press the tip of one stick at the centre of the circle to be drawn and the other tip at its rim. Rotate the compass and you will get the desired circle.[9]

It is surprising that as late as the fifteenth century, Parameśvara describes such a crude tool. Did the metal compass not reach South India or is the one described here a large version meant exclusively for drawing circles on the ground? Whatever the case may be, this

8. H. Kern, ed, *The Āryabhaṭīya* with the commentary *Bhaṭadīpikā* of Paramādīśvara [Parameśvara] (Leiden, 1874), pp. 32-33. This, in fact, is the method adopted by Sawai Jai Singh (1688-1743) in his astronomical observatories. The water channels for testing the level of the ground are still intact in Jaipur. Cf. Prahlad Singh, *Stone Observatories in India* (Varanasi, 1978), p. 75. Based on the same principle is also the levelling device called *afādain*, described by al 'Urdī in his account of the observatory at Marāgha, established in 1259. Cf. Hugo J. Seemann, *Instrumente der Sternwarte zu Marāgha nach den Mitteilungen von al 'Urdī* (Erlangen, 1928), pp. 49-50.
9. Ibid., p. 32.

account illustrates the Indian astronomer's predilection for simple make-shift tools and instruments. This point will be substantiated when one looks at Brahmagupta's choice of astronomical instruments.

II

The actual astronomical instruments described by Brahmagupta are nine in number, namely, *dhanuṣ, turyagola, cakra, yaṣṭi, śaṅku, ghaṭikā, kapāla, kartarī* and *pīṭha.*[10] Of these, *śaṅku* (gnomon) and *ghaṭikā* (clepsydra) were in use in India at least from the fourth century BC. Brahmagupta's reference to these two instruments is extremely brief, either because these two are well known instruments, or more possibly because he prefers to use other instruments in lieu of these two.

The clepsydra mentioned by Brahmagupta is the sinking-bowl type. It is a hemispherical bowl with a minute hole at the bottom and, when placed in a larger receptacle of water, it takes one *ghaṭikā* (24 minutes or 1/60th part of a civil day) to fill and sink.[11] There is

10. Bhaṭṭotpala, in his commentary on the *Bṛhatsaṃhitā*, enumerates the same nine instruments, cf. Avadha Vihari Tripathi, ed, *Bṛhatsaṃhitā*, vol. I, (Varanasi, 1966), p. 42: *yantraiś ca cakra-dhanuṣ-turyagola-yaṣṭi-śaṅku-ghaṭikā-kapāla-kartarī-pīṭhaiḥ kālaparicchedakaiḥ.* Other writers adopted the same set with minor changes. Thus Lalla's *Śiṣyadhīvṛddhida* (ch. 21) deals with twelve instruments, namely, *gola, bhagaṇa, cakra, dhanuṣ, ghaṭī, śaṅku, śakaṭa, kartarī, pīṭha, kapāla, śalākā* and *yaṣṭi.* Śrīpati's *Siddhāntaśekhara* (ch. 19) has ten: *gola, cakra, cāpa* (*dhanuṣ*), *kartārī, kapāla, pīṭha, śaṅku, ghaṭi, yaṣṭi* and *gantrī (śakaṭa).* Bhāskara's *Siddhāntaśiromaṇi (Golādhyāya, Yantrādhyāya)* discards some of Brahmagupta's instruments and introduces some of its own such as, *gola, nāḍīvalaya, ghaṭikā, śaṅku, cakra, phalaka, yaṣṭi* and *dhīyantra.*
11. *BSS*, 22.41. I-Tsing, who visited India during the lifetime of Brahmagupta, gives a detailed account of the time-keeping establishment at the Buddhist monastery of Nalanda; cf. J. Takakusu, tr., *A Record of the Buddhist Religion as Practised in India and in the Malay Archipelago* (London, 1896), pp. 144-46. While the clepsydra described by Brahmagupta (and other astronomers) measures constant *ghaṭikās* of 24 minutes each, the one at Nalanda shows variable *ghaṭikās*, that is, the length of the *ghaṭikā* varies according to the duration of the daylight. This distinction has not always been kept in mind in the little that was written on the history of horology in India.

another, and perhaps older, variety. This is the outflow type, that is, the vessel emptied itself through a hole at the base. From Varāhamihira's reference to it in the *Pañcasiddhāntikā*,[12] it appears that the vessel was a large one which emptied itself once in a nychthemeron. Jacobi thinks that such vessels must have been of cylindrical shape because it is easy to calibrate them into 60 equal units (each unit representing a *ghaṭikā*) and that the cylinder itself was called *nāḍikā* whence the time-unit received the same name.[13] Thus the basic unit of time of 24 minutes has two names in Sanskrit: *ghaṭikā* (literally, small pot), since it is measured by a small pot, that is, hemispherical bowl, and *nāḍikā* (literally, small tube) since it is ascertained by means of the cylinder-shaped outflow clepsydra. Jacobi goes on to say that the outflow clepsydra was not accurate and therefore was replaced by the sinking-bowl type. In fact, with the exception of Varāhamihira, no other astronomer mentions it among the instruments to measure time directly. Yet it survived as an essential component of complex machines.

Of the other instruments, the *cakra* (circle), *dhanuṣ* (semi-circle) and *turyagola* (quadrant) are closely related in shape and function. *Cakra* is a circular wooden plate with its circumference graduated into 360 degrees, *dhanuṣ* is its half, and *turyagola* the quarter. In all the three, a perforation is made at the centre into which a peg is inserted like an axis, and also a plumb-line is suspended from the centre. These instruments are so held towards the sun that the axis throws a shadow on the circumference. Then the arc intercepted between the nadir (indicated by the plumb-line) and the shadow is the zenith-distance. One can also measure with these instruments the angular distance between the sun and the moon, or the longitude of a planet, or the time since sunrise, and so on. Brahmagupta prefers the semi-circular variety, for he explains all functions in connection with *dhanuṣ* and adds that the same can be done with

12. *Pañcasiddhāntikā*, 14.31: *dyuniśi viniḥsṛtatoyād iṣṭacchidreṇa ṣaṣṭibhāgo yaḥ, sā nāḍī. . . .*
13. Hermann Jacobi, "Einteilung des Tages und Zeitmessung im alten Indien", *ZDMG*, vol. 74 (1920), pp. 247-63, esp. p. 251. See also J.F. Fleet, "The Ancient Indian Water-Clock", *JRAS* (1915), pp. 213-30 and F.E. Pargiter, "The Telling of Time in Ancient India", ibid., pp. 699-715.

turyagola and *cakra.*[14] But his successors show a marked preference for the *cakra*, presumably because it has an ideal shape; the *dhanuṣ* is ignored by Bhāskara while the quadrant does not appeal to Lalla or Śrīpati or Bhāskara.[15]

The other instruments are generally variations of the *cakra* or *dhanuṣ*. Thus *pīṭha* is a horizontally placed *cakra.*[16] It is a circular platform set up at the observer's eye-level, with a vertical axis equal in length to the radius of the circle. The circumference is marked with 360 degrees, and north-south and east-west lines are drawn through the centre. At sunrise and sunset, *agrās* are laid off respectively in the west and east. From the line joining the ends of these *agrās*, one can read directly the *ghaṭikās* elapsed since sunrise.[17]

The *kapāla* and *kartarī* as envisaged by Brahmagupta are variations of *dhanuṣ*. The former is a horizontally placed semi-circular plate, the axis pointing upwards, the diameter in the north-south direction, and the arc to the east or to the west where the shadow happens to fall. Like the previous instrument, this too is used for measuring the time since sunrise or the time up to sunset.[18]

In *kartarī*, two semi-circular plates are so joined that one forms the lower half of the plane of equator and the other the meridian plane, its diameter forming the polar axis. A peg is fixed at the junction of the two diameters. From its shadow on the equatorial

14. Thus Brahmagupta devotes nine stanzas for *dhanuṣ* (*BSS*, 22.8-16) and just one each for *turyagola* (17) and *cakra* (18). Āryabhaṭa also has a *dhanuṣ*, but its function is more cumbersome; cf. R.N. Rai, op. cit., p. 326. Instead of the word *dhanuṣ*, Śrīpati uses its synonyms *cāpa* (*Siddhāntaśekhara*, 19.13) and *kārmuka* (ibid., 19.17). This led some modern writers to conclude wrongly that *dhanuṣ* and *cāpa* are two different instruments; cf. S.R. Das, op. cit., pp. 262-63; D.M. Bose et al., op. cit., p. 125.
15. Cakradhara (fifteenth/sixteenth century) developed an improved model of the quadrant called *yantracintāmaṇi*; cf. R.N. Rai, op. cit., pp. 326-28.
16. *BSS*, 22.45.
17. Āryabhaṭa's *chatrayantra* works on the same principle. A similar instrument is described by Bhāskara I in his *Mahābhāskarīya,* 3.56-59, without any specific name.
18. *BSS*, 22.42-43.

plane, the time since sunrise can be read. Apparently, the equatorial plate can be rotated in its own plane, and this movement presumably gave rise to the name *kartarī* (literally, a pair of scissors).[19]

Brahmagupta's account of *yaṣṭi* (staff) is the longest of all, where he discusses several types of measurements that can be carried out with a staff, in conjunction with a plumb-line and / or a dial drawn on the ground.[20] In fact, later writers treat some of the individual functions of the staff as separate instruments. These functions are determining the time since sunrise, measuring the equinoctial shadow, finding the angular distance between the sun and the moon, estimating the heights and the distances of objects on the land, and so on. For each of these functions, Brahmagupta teaches several methods of computation.

For measuring the angular distance between the sun and the moon, two staves are used. These are joined together at the lower end, the upper ends pointing to the sun and the moon respectively. In this position, the two staves look like the Roman letter "V" or like two shafts of a cart joined at the yoke. Hence Lalla and Śrīpati treat this as a separate instrument called *śakaṭa* (cart).[21] When the staff is used in land survey for measuring, for instance, the height and distance of buildings, this particular function is designated as *dhīyantra* by Bhāskara because it is the *dhī* (intellect) of the observer that is important.[22]

These, in brief, are the instruments and their functions according to Brahmagupta. Several of these are mentioned by him for the first time, but it is immaterial whether he invented any of them. More important is that he offers the first systematic study of the instruments and their use, with several alternative methods of computation, to go into which is beyond the scope of this paper. But when we compare the instruments selected by Brahmagupta with those mentioned by Āryabhaṭa or Varāhamihira, we are struck by the utter simplicity of design in Brahmagupta's instruments.

19. *BSS*, 22.43-44. In their version of this instrument, Lalla (*Śiṣyadhīvṛddhida*, 21.24) and Śrīpati (*Siddhāntaśekhara*, 19.14) use only a single semi-circular plate to represent the equatorial plane.
20. *BSS*, 22.19-38.
21. *Śiṣyadhīvṛddhida*, 21.42-47; *Siddhāntaśekhara*, 19.26-27.
22. *Siddhāntaśiromaṇi, Golādhyāya, Yantrādhyāya*, 40-49.

Leaving aside the clepsydra—since there were many alternative methods of measuring the time since sunrise, it was not really essential—the rest of the instruments can be reduced to two basic types: a wooden circle with a graduated rim, and a wooden staff. With these two, the astronomer measured the zenith distance as an arc and the gnomonic shadow as a line, and calculated the rest from such simple measurements.

These instruments, and also the accessories mentioned before, are such that they can be manufactured everywhere with little or no skill. The basic idea seems to be that a tool or an instrument is just a means to an end. The end may be elaborate but the means must be the simplest possible. Even Bhāskara II who invented the *phalaka-yantra,* the only pre-Islamic instrument of some complexity, shares this attitude when he declares:

> What does a man of genius want with complex instruments on which scores of books have been written? Let him just take a staff in his hand and cast his eye from its one end to the top. There is no object in sight of which he cannot find out the measurements, be it on the earth, in water, or in the sky.[23]

This does not mean that the Indian astronomer totally ignored the observational aspect. He was well aware that the success of the *siddhānta* depended upon *dṛksāmya*, that is, the exact correspondence between the computed and observed results. When there was none, the *siddhānta* had to be revised accordingly. Thus Nīlakaṇṭha says: *yaḥ siddhānto darśanāvisaṃvādi bhavati so 'nveṣaṇīyaḥ:* "where there are many textbooks, choose one that accords well with actual observations".[24] But it was thought that a few occasional sightings were enough for this purpose, and we have no record of systematic observations stretching over a long span of time (except in the case of Parameśvara of Kerala, who computed and observed a large number of eclipses between AD 1393 and 1432).

The result was that astronomical instruments did not develop any further from their simple design, nor did they lead to any significant advances in instrumentation technology. The instruments were of

23. Ibid., 40.
24. In the *Jyotirmīmāṃsā,* quoted in *Source-Book*, p. 7.

the simplest design and were made of wood or bamboo. These, then, could not have provided much precision in measurement, nor did their construction involve a meaningful interaction between the *siddhāntin* and *śilpin*. Hence these instruments could not vie with the astrolabe when it was introduced into India in the thirteenth century. On this versatile instrument, several elements can be read off directly without resorting to lengthy computations. Since its manufacture required a high degree of cooperation between the astronomer and the artisan, it also offered greater precision. Consequently, when Sawai Jai Singh II (1686-1743) looked for prototype instruments for carrying out systematic and precise observations, Sanskrit tradition had nothing to offer him, and he had to turn to Central Asian models.

III

The picture of the Indian astronomer that emerges from the discussion so far is that of one who took an occasional observation with the simplest of instruments but depended otherwise on his superior computational skill. No special tool or material was needed for computation either; it was carried out with the tip of his finger writing in the dust and was even called *dhūlikarman*. Therefore, it comes as a surprise that Brahmagupta and others devoted considerable space to complex automatic devices called *svayaṃvaha yantras* (self-propelling machines). It is difficult to reconcile these two attitudes: on the one hand, stark austerity as regards what we would consider today as an astronomer's essential equipment; on the other, free rein to imagination in devising more and more complex or amusing automata. Possibly, to a people forced to empty the sinking-bowl type of clepsydra sixty times in the course of a nychthemeron, the notion of a *yantra* that turned by itself was greatly tempting. Be that as it may, these devices form an interesting chapter in the history of human effort to master the sources of energy, and Brahmagupta's contribution to this chapter is not insignificant. It is, however, convenient to begin with Āryabhaṭa because he was the first to describe such machines. In a cryptic stanza in his *Āryabhaṭīya*, he says:

The Sphere which is made of wood, perfectly spherical, uniformly dense all round but light (in weight) should be made to rotate keeping pace with time with the help of mercury, oil and water by the application of one's own intellect.[25]

Here he is obviously referring to a globe that rotates around its axis automatically at the rate of one rotation per 24 hours, but he does not tell us how to construct such a device. Fortunately, his commentators, Someśvara (eleventh/twelfth century), Sūryadeva Yajvan (b. AD 1191), Parameśvara (*c.* AD 1450) and Nīlakaṇṭha Somasutvan (*c.* AD 1501), explain the process with a rare degree of unanimity. This is how Sūryadeva Yajvan describes the apparatus:

Having set up two pillars on the ground, one towards the south and the other towards the north, mount on them the ends of the iron needle (rod) (which forms the axis of rotation of the Sphere). In the holes of the Sphere, at the south and north poles, pour some oil so that the Sphere may rotate smoothly. Then, underneath the west point of the Sphere, dig a pit and put into it a cylindrical jar with a hole in the bottom and as deep as the circumference of the Sphere. Fill it with water. Then, having fixed a nail at the west point of the Sphere, and having fastened one end of a string to it, carry the string downwards along the equator towards the east point, then stretch it upwards and carry it to the west point (again), and then fasten it to a dry hollow gourd (appropriately) filled with mercury and place it on the surface of water inside the cylindrical jar underneath, which is already filled with water. Then open the hole at the bottom of the jar so that with the outflow of water, the water inside the jar goes down. Consequently, the gourd which, due to the weight of mercury within it, does not leave the water, pulls the Sphere westwards. The outflow of water should be manipulated in such a way that in 30 *ghaṭīs* (= 12 hours) half the water of the jar flows out and the Sphere makes one half of a rotation, and similarly, in the next 30 *ghaṭīs* the entire water of the jar flows out, the gourd reaches the bottom of the jar and the Sphere performs one complete rotation.[26]

25. *Āryabhaṭīya, Gola,* 22: *kāṣṭhamayaṃ samavṛttaṃ samantataḥ samaguruṃ laghuṃ golam, pāradatailajalais taṃ bhramayet svadhiyā ca kālasamam.* The translation is by K.S. Shukla; cf. his edited and translated version of the *Āryabhaṭīya*, op. cit., p. 129.
26. See K.V. Sarma, ed, *Āryabhaṭīya* with the commentary by Sūryadeva Yajvan (New Delhi, 1976), pp. 143-44 for the Sanskrit text; and K.S. Shukla, ed & tr, *Āryabhaṭīya*, op. cit., pp. 129-30 for the English translation.

In other words, this apparatus is powered by an outflow type clepsydra which empties itself in 60 *ghaṭikās*. Varāhamihira prescribed such a clepsydra for measuring time directly, but other astronomers rejected it. In his *Āryabhaṭasiddhānta* also, Āryabhaṭa describes some devices powered by the same type of cylindrical clepsydra which he calls a hollow pillar.[27]

Brahmagupta first makes an innovation in this clepsydra. He suggests that the length of the cylindrical jar (*nalaka*) should be calibrated into 60 equal divisions, each one denoting a *ghaṭikā*. Then, instead of simple string, a long narrow strip of cloth (*cīrī*) should be attached to the mercury-filled gourd. In this strip of cloth, 60 knots are tied at distances equal to the divisions marked on the cylinder, and the knots are numbered serially. Then, as the float goes down it pulls the strip of cloth with the knots downwards, and the passage of each knot beyond a certain point indicates the passage of a *ghaṭikā*.[28]

With this basic design, Brahmagupta devises a number of models. In the first, a male doll is set up in such a manner that, as the float goes down, the strip of cloth issues out of the doll's mouth. Thus after each *ghaṭikā*, the doll "spits out" a knot with the appropriate serial number written on it.[29] Or one can set up two dolls, a bride and bridegroom, and the numbered knots pass from the bridegroom's mouth to the bride's, like sweetmeat in some marriage ritual.[30] Another variation is to fix a small figurine or jack on each knot. As it passes a certain point, it releases a lever which hits a drum or rings a bell. Thus, after each *ghaṭikā*, a drum can be beaten or a bell rung.[31] Likewise, a peacock can be so constructed that it swallows or vomits a certain length of a snake in each *ghaṭikā*.[32]

All these ingenious devices are based on the erroneous assumption that the water level in the clepsydra falls by equal distances in equal time intervals. But the clepsydra used here is a

27. See *Source-Book*, p. 84.
28. *BSS*, 22.46-47.
29. Ibid., 22.47-48.
30. Ibid., 22.50.
31. Ibid., 22.51-52.
32. Ibid., 22.51.

regular cylinder called *stambha* by Āryabhaṭa and *nalaka* by Brahmagupta, and here the outflow of water cannot be uniform because of the changing water pressure. Consequently, the *ghaṭikās* indicated by these devices will not be of the same duration.

Nīlakaṇṭha Somasutvan raises this problem for the first time in his commentary on the *Āryabhaṭīya* written at the beginning of the sixteenth century. He says that if the cylinder has the same circumference at the top and the bottom, the outflow will be faster at the beginning, with the result that the sphere will make a quarter rotation long before it is midday. Nīlakaṇṭha's solution to this problem is to vary the circumference at the top and the bottom, but he does not say how this should be done.[33]

Nīlakaṇṭha's solution is not entirely correct nor is it novel. The ancient Egyptians tried to regulate the water pressure by adopting a vessel with sloping sides. Their clepsydras shaped like truncated cones or buckets are attested from the thirteenth century BC, but these did not show uniform time intervals either.[34] Therefore, in the Classical world, the Romans introduced cylindrical inflow clepsydra. Here water dripped into the cylindrical vessels from a reservoir in which the water level was kept constant. Hence, water dripped into the cylinder with a uniform speed and the water level in the cylinder rose accordingly. A float was set up in the cylinder, which rose along with the water level and marked the time against a scale. The same principle was followed in China.

Coming back to the automatic devices of Āryabhaṭa and Brahmagupta, these are survivals from the period when outflow clepsydra was used for measuring time directly. These automatic devices are technically feasible, but would indicate irregular time intervals. Lalla and Śrīpati accept Brahmaguta's automata without

33. *The Āryabhaṭīyam* with the Bhāṣya of Nīlakaṇṭha Somasutvan, part III (Trivandrum, 1957), pp. 38-39.
34. Curiously enough, in the twelfth century AD, al-Jazarī employs a bucket-shaped clepsydra with a float as the driving force for one of his water clocks. In this connection, Donald Hill remarks: "It would be exceedingly difficult to make a vessel by empirical methods, so that the water level fell by equal distances in equal time intervals". See Ibn al-Razzāz al-Jazarī, *The Book of Knowledge of Ingenious Mechanical Devices*, tr. and annotated by Donald R. Hill (Dordrecht, 1974), pp. 71-74, 250-51, esp. p. 250.

any hesitation,[35] but Bhāskara II rejects them as rustic contrivances (*grāmya*). His reason for rejection is not the absence of a uniform outflow—he is silent on this point—but that the cylinder has to be filled afresh every day. He would like instruments that turn absolutely on their own without the aid of any human agency (*nirapekṣa*) and for ever.[36] The credit for conceiving such a machine goes to Brahmagupta.

IV

Brahmagupta's *perpetuum mobile* or machine with perpetual motion is a wheel made of light wood. Into its rim are inserted, at equal intervals, hollow spokes of equal size. Each spoke is half filled with mercury and then sealed. When the axle of this wheel is set up on two supports, the mercury runs up and down the spokes, and the wheel turns perpetually (*ajasraṃ bhramati*).[37] But what is the connection between this wheel and astronomy? Brahmagupta seems to think that by regulating the quantity of mercury, the speed of rotation can be so adjusted that the wheel functions as a time-keeping device.[38]

The idea of a mercury-powered wheel with perpetual motion seems to be Brahmagupta's own. Just as the alchemist thought that mercury can transmute base metals into gold, so does Brahmagupta hold that it can overcome inertia and cause the wheel to turn eternally. The belief in the miraculous power of mercury persisted in the eleventh century also when King Bhoja of Dhārā assumed that it could overcome gravity and so raise an aerial car from the ground.[39]

35. *Śiṣyadhīvṛddhida*, 21.10-17; *Siddhāntaśekhara*, 19.7-11.
36. *Siddhāntaśiromaṇi, Golādhyāya, Yantrādhyāya*, 57.
37. *BSS*, 22.53-54.
38. *BSS*, 22.55.
39. *Samarāṅgaṇa-Sūtradhāra*, ed, T. Ganapati Sastri and revised by V.S. Agrawala (Baroda, 1966), 31.95-100. Some time-keeping devices are also described at 31.66-71, but their mechanism is obscure. See also V. Raghavan, *Yantras or Mechanical Contrivances in Ancient India* (Indian Institute of Culture, Transaction No. 10), 2nd edn (Bangalore, 1956), pp. 24, 29.

Brahmagupta's *perpetuum mobile* was taken up enthusiastically by Lalla and Bhāskara. The former states that if the wheel with mercury-filled spokes is joined to the axle of an armillary sphere, it will rotate the armillary sphere continuously.[40] Bhāskara II suggests that the spokes should be slightly curved, all in the same direction as in a *nandyāvarta*. The wheel will then turn for ever because at some places the mercury runs towards the nave of the wheel and at other places towards the rim. He also proposes a new variation, in which a channel is cut in the rim of the wheel and filled half with water and half with mercury.[41]

One is apt to ridicule these devices as mere flights of fantasy and Brahmagupta's treatment of instruments as a tiresome mixture of science and superstition. Even Bhāskara did not see any connection between the serious pursuit of astronomy and these self-propelling devices. For him these were part of the juggler's (*kuhaka*) equipment, and he discussed them only because the previous astronomers (like Brahmagupta) had done so.[42] But he does not seem to entertain any doubts about the functioning of these devices.

Lynn White argues that such fantasies are also important in the history of ideas, and that the notion of perpetual motion, originating in India, led to important inventions in Europe. He observes:

> The symptom of the emergence of a conscious and generalized lust for natural energy and its application to human purposes is the enthusiastic adoption by thirteenth-century Europe of an idea which had originated in twelfth-century India—perpetual motion.[43]

Lynn White goes on to describe Bhāskara's two versions of *perpetuum mobile* and shows how this idea was instantly picked up and elaborated upon by the Islamic World and then transmitted to Europe. There the idea was received with intense and widespread interest and soon attempts were made to apply it for the benefit of mankind. Thus was laid the foundation of power technology in the modern world.

40. *Śiṣyadhīvṛddhida*, 21.18-19.
41. *Siddhāntaśiromaṇi, Golādhyāya, Yantrādhyāya*, 50-53.
42. Ibid., 58.
43. L. White, *Medieval Technology and Social Change* (Oxford, 1962), pp. 129-30.

Now this study of Brahmagupta's instruments should enable us to see that the idea of perpetual motion, or more precisely the design for a perpetually moving machine, originated not in twelfth century India but much earlier—with Brahmagupta in the seventh century.

Needham, however, detects Chinese inspiration in Indian devices. He states:

> One gets a strong impression from some of the Sanskrit texts that the writer was trying to describe water-wheel clocks of Chinese type in veiled language, or else that he knew only vaguely how they worked. Indeed one begins to entertain the belief that the stimulus for the flood of ideas on perpetual motion devices may have been derived from Indian monks or Arabic merchants standing before a clock-tower such as that of Su Sung and marvelling at its regular action.[44]

In the light of the evidence which we have, it is difficult to agree with Needham's line of thinking. The Sanskrit texts he has in mind are the *Sūryasiddhānta* and the *Siddhāntaśiromaṇi* of Bhāskara; he is not aware that it is Brahmagupta who realized the feasibility of a perpetual motion machine. Moreover, Su Sung's clock-tower was built in AD 1090 and received the power from a water-wheel that was made to turn in the following manner:

> Water stored in the upper reservoir is delivered into the constant-level tank by a siphon and so passes to the scoops of the driving-wheel. . . . As each scoop in turn descends the water is delivered into a sump. Apparently the clock was never so located as to be able to take advantage of a continuous water supply; instead of this, the water was raised by hand-operated norias in two stages to the upper reservoir.[45]

On the other hand, our attempt in this paper to chronologically study the automatic devices from Āryabhaṭa up to Bhāskara establishes, it is hoped, that these devices had a long history in India and operated on a totally different principle, namely, outflow clepsydra and sinking float, which was never popular in China as Needham admits.[46]

44. J. Needham, *Science and Civilisation in China*, vol. IV, part II (Cambridge, 1965), p. 540.
45. Ibid., pp. 457-58.
46. Ibid., p. 469.

However, Su Sung's clock-tower found an echo in India in one instance. Besides the two varieties of mercury-powered *perpetuum mobile* which we have discussed above, Bhāskara describes a third one. This consists of a large wheel to the rim of which pots are attached as in a noria. Through a copper siphon water is released into the topmost pot from a reservoir. As the pot is filled, it becomes heavy and goes down, and thus the wheel is rotated. The water falling down from the wheel is collected in a channel and "is made to go up into the reservoir, so that it need not be filled again".[47] This is the first and only mention of such water-wheels in Sanskrit texts. Bhāskara describes the siphon and its working principle in great detail as if it were a novelty. It would be interesting to investigate what Bhāskara's sources are.

After Bhāskara, the idea of perpetual motion does not seem to have enthused any astronomer in India. In the early seventeenth century, Raṅganātha thought that such machines were possible only in Europe. Commenting on the *Sūryasiddhānta's* reference to these devices, he declares:

> This science of self-propelling machines (*svayaṃvaha-vidyā*) is well practised by the people known as phiraṅgīs who live beyond the Seven Seas. Since it is part of jugglery (*kuhaka-vidyā*), there is no need to discuss it here at length.[48]

Raṅganātha belonged to an influential family of astrologers with contacts at the Moghul court. His brother was a protégé of Khān-i-Khānān 'Abd al-Rahīm Khān and a favourite of Jahāngīr.[49] Raṅganātha, therefore, may have seen or heard about the mechanical clocks and other contrivances brought by Europeans as gifts to the Moghul court. His, then, is one of the first Indian responses to European technology.

47. *Siddhāntaśiromaṇi, Golādhyāya, Yantrādhyāya,* 53-56.
48. In his commentary called *Gūḍhārthaprakāśa* on *Sūryasiddhānta* 13. 22. The commentary is available in several editions.
49. For his genealogy, see Pingree, op. cit., p. 126.

3

Perpetual Motion Machines and their Design in Ancient India

Histories of technology are perforce Eurocentric, not so much because of racial prejudice but owing to the paucity of the detailed spadework that needs to precede the writing of comprehensive account of technologies outside Europe. Joseph Needham's monumental effort in unravelling the *Science and Civilization in China* forms an exception to this state of affairs. Significant likewise are also the studies on medieval technology by Lynn White Jr., who invites attention to the concepts and inventions of other Asian nations. Notable in this connection is his seminal essay *Tibet, India, and Malaya as Sources of Western Medieval Technology*, published some thirty years ago.[1] One of the concepts whose origin he attributes to India is the perpetual motion machine. For students of history of technology in India it will be instructive, both conceptually and methodologically, to take a closer look at Lynn White's thesis and the controversy it engendered.

A perpetual motion machine (Latin *perpetuum mobile*, Sanskrit *ajasra-yantra*) is a device that is supposed to perform useful work without any external source of energy or, at least, where the output is far greater than the input. The idea of constructing such machines and of employing the power generated by them has fascinated the

First published in: *PHYSIS: Rivista Internationale di storia della Scienza*, 29.3 (1992) 665-76.

1. L. White Jr., *Tibet, India, and Malaya as Sources of Western Medieval Technology,* "American Historical Review", LXV, 1960, pp. 515-526; reprinted in Id., *Medieval Religion and Technology. Collected Essays,* Berkeley, 1978, pp. 43-57. See also, Id., *Medieval Technology and Social Change,* London, 1964, pp. 129-131.

Fig. 3.1. Perpetual motion wheel according to Brahmagupta.

minds of many inventors in Europe since the Middle Ages.[2] Modern science says that it is impossible to construct such machines and ridicules the attempts as mere flights of fantasy. Lynn White, however, argues that such fantasies are also important in the history of ideas and that the concept of perpetual motion was a significant element in Europe's thinking about mechanical power.

White traces the origin of the perpetual motion machine to twelfth-century India; in particular to Bhāskara, the great mathematician and astronomer, who in his *Siddhāntaśiromaṇi* (AD 1150) describes two wheels which are supposed to turn for ever. In the first model (Fig. 3.2) the hollow spokes are half filled with mercury and in the second (Fig. 3.3) a narrow channel is scooped out in the rim and filled half with mercury and half with water. According to White, these two models were immediately taken up by the Islamic World and amplified. The amplifications

2. In India, Sawai Jai Singh is reported to have invested a fortune in constructing one. Father Andreas Strobl, the Bavarian Jesuit astronomer who was at Jai Singh's court in the 1740s, reports in a letter dated 18 October 1743 that at the time of his death Jai Singh was busy erecting a machine which he had invented to demonstrate perpetual motion and that he had already spent 50,000 guilders on this venture. Cf. S. Noti, *Land und Volk des koeniglichen Astronomen Dschaisingh II Maharadscha von Dschaipur*, Berlin, 1911, p. 98.

Fig. 3.2. Bhāskara's first model of perpetual motion wheel.

can be seen in an anonymous Arabic manuscript which contains the designs for six perpetual motion wheels, and one of these (Fig. 3.4) closely resembles Bhāskara's first model with mercury-filled spokes.[3] The Islamic World in turn transmitted the idea to the West at the beginning of the thirteenth century, together with Indian numerals and the decimal place-value system.

Europe responded to this idea of *perpetuum mobile* with great enthusiasm, engineers like Villard de Honnecourt and Peter of Maricourt designing several new models. Two of the six wheels found in the Arabic manuscript reappear in Villard's notebooks and Bhāskara's second model does so in an anonymous Latin work of the late fourteenth century. In contrast to India and the Islamic World, the medieval engineers of Europe tried to apply the idea of perpetual motion for practical purposes, for the material benefit of mankind. Already the industrial application of water- and wind-power was revolutionizing manufacture, and the two new forces introduced by the Islamic World, viz., gravity and magnetism, appeared to operate with a constancy unrivalled by wind or water. Lynn White describes the quest for energy by these thirteenth-century European engineers thus:

3. The resemblance was first noticed by H. Schmeller, *Beiträge zur Geschichte der Technik in der Antike und bei den Arabern,* Erlangen, 1922, pp. 16-19.

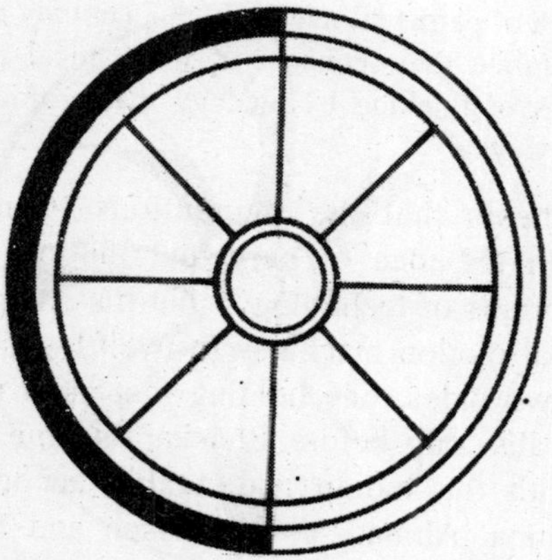

Fig. 3.3. Bhāskara's second model of perpetual motion wheel.

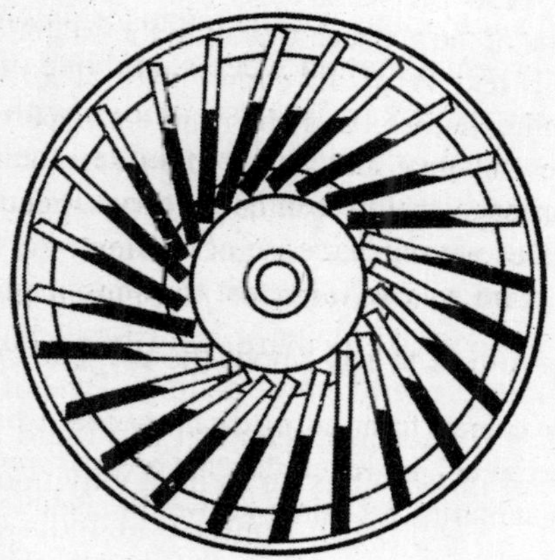

Fig. 3.4. Perpetual motion wheel from an Arabic manuscript.

They were coming to think of the cosmos as a vast reservoir of energies to be tapped and used according to human intentions. They were power-conscious to the point of fantasy. But without such fantasy, such soaring imagination, the power technology of the Western world would not have been developed.[4]

He concludes his thesis by saying:

4. L. White Jr., *Medieval Technology,* op. cit., p. 134.

Thus the Indian idea of perpetual motion [. . .] not only helped European engineers to generalize their concept of mechanical power, but also provoked a process of thinking by analogy that profoundly influenced Western scientific views.[5]

Lynn White's thesis that the foundations of modern power technology lay in the idea of perpetual motion was generally accepted by historians of technology,[6] but his attempt to trace the origin of perpetual motion machines to twelfth-century India was contested from two sides, one holding that such machines were known to the Arabs long before Bhāskara's time, and the other claiming that both the Indian and Arabic accounts owe their inspiration to China. Ahmad Y. Al-Hassan and Donald R. Hill argue in their excellent book *Islamic Technology; An Illustrated History* thus:

In India about AD 1150 Bhāskara described a perpetual motion wheel which resembles one of the six such wheels in the Arabic manuscripts, but the original Arabic text is of an earlier date. The Arabic technical descriptions, the illustrations, and the whole complex of the sixteen machines are quite elaborate and, as we have seen, constitute a single approach. The occurrence, therefore of one or two perpetual-motion wheels in the Indian text does not imply a case of transmission from one culture to another, though there was an important transmission to the West.[7]

The authors have this to say about the original Arabic text:

It must have been copied from an original treatise which is at present unknown to us. We can tell, however, that this original was written between the third and sixth centuries AH (ninth to twelfth centuries AD).[8]

5. Id., *Medieval Religion*, op. cit., pp. 56-57.
6. J. Needham, *Science and Civilization in China*, vol. IV, part 2, Cambridge, 1965, p. 54: "Lynn White has done a good service by pointing out that in correct historical perspective, the idea of perpetual motion has heuristic value". See also J. Gimpel, *The Medieval Machine. The Industrial Revolution of the Middle Ages,* Aldershof, 1988 (2nd edn.), pp. 127-129; R. Lannoy, *The Speaking Tree. A Study of Indian Culture and Society*, London, 1971, pp. 290-293.
7. A.Y. Al-Hassan, & D.R. Hill, *Islamic Technology. An Illustrated History,* Cambridge, 1985, p. 71.
8. Ibid., p. 70.

Even though they are rather vague about the original Arabic text, it will be shown presently that the Indian concept of the *perpetuum mobile* is much older than Bhāskara and also older than the alleged antiquity of the "unknown and undated" original Arabic text.

The second party of opposition to Lynn White's view is represented by Joseph Needham, who asserts that

> Indeed one begins to entertain the belief that the stimulus for the flood of ideas on the perpetual motion devices may have been derived from Indian monks or Arabic merchants standing before a clock tower such as that of Su Sung and marvelling at its regular action.[9]

Lynn White dismissed this suggestion as lacking in any evidence.[10]

The astonishing thing about this debate—like many other debates concerning India's past—is that it is conducted on the basis of just two Sanskrit texts which happen to be available in English translation, ignoring all other texts. Lynn White traces the idea of the *perpetuum mobile* to twelfth-century India on the basis of Lancelot Wilkinson's translation of the *Siddhāntaśiromaṇi,*[11] while Needham's comments emanate from his perusal of Ebenezer Burgess's rendering of the *Sūryasiddhānta.*[12] The passage cited by Needham does not even discuss the *perpetuum mobile*. No doubt, Lynn White's conclusions are highly perceptive even with the limited sources available to him, but in history of technology there are no shortcuts. One has to study all the relevant original texts, and the material remains if there are any, and interpret the data in the correct space-time framework. In the present case, a study of the original texts not only upholds Lynn White's view, but even strengthens it further. Before we discuss the evidence of the Sanskrit sources, one distinction has to be made.

9. J. Needham, op. cit., p. 540.
10. L. White Jr., *Medieval Religion,* op. cit., p. 53, n. 60 (= Id., *Medieval Technology,* op. cit., p. 130, n. 3).
11. For the text, see Bhāskara, *Siddhāntaśiromaṇi*, ed. Bapu Deva Sastri, rev. Ganpati Deva Sastri, Benares, 1920, Golādhyāya, Yantrādhyāya, vv. 50-53.
12. For the text, see *The Sūryasiddhānta*, with the Exposition of Ranganātha, the Gūdhārtha-prakāśika, ed. F. Hall, reprint, Amsterdam, 1974, 13.16-18.

The Sanskrit astronomical texts describe two kinds of automatic devices, both called *svayaṃvaha yantra,* "self-propelled machines". In the first variety, an outflow type of water clock causes a solid sphere to rotate around its axis once in 24 hours, thus simulating the apparent motion of the great circles in the heavens. This teaching instrument was described for the first time by Āryabhaṭa about the beginning of the sixth century AD.[13] This apparatus is naturally dependent on human agency to replenish the water regularly. The second variety is the *perpetuum mobile* that is supposed to turn for ever without any external help. Some Sanskrit texts designate this one as *ajasra-yantra* "perpetual machine" and it is with this variety that we are concerned here.[14]

This device was described for the first time, not by Bhāskara in the middle of the twelfth century as Lynn White supposed but more than half a millenium before him, by Brahmagupta, another great mathematician and astronomer. His *Brāhmasphuṭasiddhānta,* completed in AD 628, contains the first systematic treatment of the construction and use of a large number of scientific instruments.[15] Here Brahmagupta describes the *perpetuum mobile* thus (Fig. 3.1):

> Make a wheel of light timber, with uniformly hollow spokes at equal intervals. Fill each spoke up to half with mercury and seal its opening situated in the rim. Set up the wheel so that its axle rests horizontally on two [upright] supports. Then the mercury runs upwards [in some] hollow spaces and downwards [in some others, as a result of which] the wheel rotates automatically for ever.[16]

13. S.R. Sarma, *Astronomical Instruments in Brahmagupta's Brāhmasphuṭasiddhānta,* "Indian Historical Review", XIII, 1986-87, pp. 63-74, esp. 69-71; reprinted in this Volume, pp. 47-63.
14. Bhāskara, *Siddhāntaśiromaṇi*, Golādhyāya, Yantrādhyāya, v. 57, declares that the former are rustic (*grāmya*) contraptions as they require the help of human agency, while the real ingenuity (*yukti*) lies in the latter variety.
15. S.R. Sarma, op. cit.
16. Brahmagupta, *Brāhmasphuṭasiddhānta*, ed. Sudhakara Dvivedi, Benares, 1902, 22.53-54:

laghudārumayaṃ cakraṃ samasuṣirārāntaraṃ pṛthag arāṇām /
ardhe rasena pūrṇe paridhau saṃśliṣṭakṛtasandhiḥ //
tiryak kīlo madhye dvyādhārastho 'sya pārado bhramati /
chidrāṇy ūrdhvam adho 'taś cakram ajasraṃ svayaṃ bhramati //

What is the purpose of this wheel in the midst of instruments meant for measuring fractions of the day time or the solar or stellar altitudes? In Indian philosophical thinking, time is conceived in two ways: *khaṇḍa-kāla*, measurable time, and *akhaṇḍa-kāla,* immeasurable, eternal and cyclical Time.[17] While the other instruments are meant to measure the *khaṇḍa-kāla*, Brahmagupta obviously wishes to demonstrate the eternal and cyclical flow of the *akhaṇḍa-kāla* through his quicksilver wheel which, he believes, turns for ever and ever.

Lynn White and Needham note that this concept is quite in consonance with the "Hindu concept of cyclical and self-renewing nature of all things",[18] or with the concept of "*kalpas* and *maha-kalpas* succeeding one another in self-sufficient and unvarying round".[19] But between the philosophical notion of the eternally moving cyclical Time on the one hand and the technical design for the perpetual motion wheel on the other, there is a crucial intermediate step which needs to be emphasized. This step is represented by the concept of the universe as a perpetually turning wheel. There are abundant examples of Indian thought to illustrate this concept. The *Bhagavadgītā* (18.61), for instance, states that the Lord causes all things to revolve as though they are mounted on a wheel. Buddhism conceives the regulating principle of the universe as a wheel (*dhamma-cakka*) which, when once set in motion by the Buddha, goes on and on. It is this concept of the universe as an eternally rotating wheel that inspired Brahmagupta to design the quicksilver wheel which, in the words of Richard Lannoy, "is a uniquely Indian product of the aesthetically oriented creative imagination which seeks to harness and balance the forces of nature".[20] But it has larger historical consequences as well.

It may be recalled that Lynn White asserts that Bhāskara's mercury wheels were instantly picked up by the Islamic World, and that Al-Hassan and Hill counter this by saying that the original Arabic work on these wheels predates Bhāskara. In fact, Lynn

17. See, for example, *Sūryasiddhānta* 1.10.
18. L. White Jr., *Medieval Technology*, op. cit., p. 130.
19. J. Needham, op. cit., p. 540.
20. R. Lannoy, op. cit., pp. 291-292.

White's linkage has no foundation because Bhāskara's work is not known to have been translated into Arabic or otherwise transmitted to the Islamic World.

On the other hand, it is quite well known that Brahmagupta's works were so transmitted in the second half of the eighth century.[21] Under the Caliphate of al-Mansur (754-775), there was a great flow of scientific ideas from India to Baghdad. The astronomical works of Āryabhaṭa and Brahmagupta, Indian numerals with decimal place-value, and the like were transmitted to, and adopted by, the Islamic World. Based on the Arabic adaptations of Brahmagupta's *Brāhmasphuṭasiddhānta*, there developed the Sindhind school of astronomy. Likewise, certain basic methods of observational astronomy were also borrowed by the Arabs from India. For example, the method taught in the Sanskrit texts for determining the cardinal directions with the shadow of a gnomon came to be known among the Arabs as the "Indian circle". Mu'ayyad al-Dīn al 'Urḍī, the celebrated instrument maker at the Marāgha Observatory (established in 1259), states that the Indian circle is the best among the innumerable methods of finding the cardinal directions.[22] His technique of aligning a plane horizontally by means of water has antecedents in India.[23]

In view of these transmissions, it is quite possible that the Arabs became acquainted, in the latter half of the eighth century, with the idea of the *perpetuum mobile* as described in the *Brāhmasphuṭasiddhānta* which was available to them in Arabic. Al-Hassan and Hill, it will be recalled, assign the original Arabic text containing the descriptions of *perpetua mobilia* to a period between the ninth and the twelfth centuries, without clearly stating the reasons. Even if the earliest date, viz., ninth century, is accepted,

21. On the transmission of Indian astronomy to the Islamic World, see F. Sezgin, *Geschichte des Arabischen Schrifttums*, vol. VI, Leiden, 1978, pp. 116-120; M. Mohd. Yousuf, *Influence of Indian Sciences on Muslim Culture*, "Islamic Culture", XXXVI, 1962, pp. 102-118.
22. H.J. Seemann, *Die Instrumente der Sternwarte zu Marâgha nach den Mitteilungen von al 'Urḍī*, Erlangen, 1928, pp. 24-25; E. Wiedemann, *Ueber den indischen Kreis,* "Mitteilungen zur Geschichte der Medizin und Naturwissenschaften", XI, 1912, pp. 252-255.
23. H.J. Seemann, op. cit., pp. 49-50; S.R. Sarma, op. cit., p. 65; see in this Volume, pp. 49-50.

Brahmagupta's mercury wheel reached the Arabs at least one century before that date.

Brahmagupta's mercury wheel is earlier also than Su Sung's clock tower (AD 1090)[24] and other Chinese devices. I-Tsing, a Buddhist traveller who came to India during Brahmagupta's life time, showed great appreciation of Indian water clocks but he himself was not aware that water clocks (though of different type) were also available in China.[25] Therefore, the question of his, or an Indian monk's, imparting the knowledge of Chinese automatic clocks to Indian astronomers like Brahmagupta does not arise, contrary to what Needham would like to believe.

Thus perpetual motion machines, numerals in decimal place-value system, sine tables, etc., are notable elements in a complex set of ideas that were transmitted from India to the Islamic World in the eighth century and thence to the West at about the beginning of the thirteenth century. If the idea of perpetual motion played a role in the development of mechanical power in Europe, the source then is not Bhāskara of the twelfth century but Brahmagupta of the seventh.

However, Bhāskara did make some important innovations in mercury wheels. As has been stated already, his *Siddhāntaśiromaṇi* contains the description of two mercury wheels. The first one is similar to Brahmagupta's, even in the wording of the description.[26] But Bhāskara adds that the hollow spokes should be *kiñcid vakrāḥ*. Wilkinson rendered this expression as "let them also be all placed at an angle somewhat verging from the perpendicular",[27] and this

24. J. Needham, W. Ling, D. J. De Solla Price, *Heavenly Clockwork. The Great Astronomical Clocks of Medieval China,* Cambridge, 1986 (2nd edn.).
25. I-Tsing, *A Record of the Buddhist Religion as Practised in India and the Malay Archipelago,* Engl. trans. J. Takakusu, reprint Delhi, 1966, p. 146; S.R. Sarma, *Water Clocks and Time Measurement in India,* forthcoming.
26. Bhāskara, *Siddhāntaśiromaṇi,* Golādhyāya, Yantrādhyāya, vv. 50-51 ab:

    laghudārujasamacakre samasuṣirārāḥ samāntarā nemyām /
    kiñcid vakrā yojyāḥ suṣirasyārdhe pṛthak tāsām //
    rasapūrṇe tac cakraṃ dvyādhārākṣasthitaṃ svayaṃ bhramati /
27. As quoted by L. White Jr., *Medieval Technology*, op. cit., p. 130.

translation led Lynn White to conclude that the spokes are slanted as in one of the wheels (Fig. 3.4) of the Arabic manuscript.

In his commentary, Bhāskara himself explains that *kiñcid vakrāḥ* means that the spokes should be slightly curved, all towards the same direction like the petals of the *nandyāvarta* flower (*Tabernamontana coronaria)* (Fig. 3.2). Because of such curvature, Bhāskara goes on to explain, the mercury in one part of the wheel runs fast towards the bottom of the spokes while in another part it runs towards the top of the spokes. Impelled by this internal movement, the wheel itself turns automatically and will continue to do so.[28]

Besides thus modifying Brahmagupta's wheel, Bhāskara designed another variant in which a narrow channel is cut in the surface of the rim and filled half with mercury and half with water and then sealed up (Fig. 3.3). The water, trying to flow downwards, pushes the mercury and vice versa, this internal tension resulting in the rotation of the wheel itself.[29]

The last word on automata was uttered in India by Raṅganātha who, writing in 1603, observed that the Europeans were great experts in the science of automata.[30] He belonged to an influential

28. Bhāskara, *Siddhāntaśiromaṇi*, p. 248: *tāś ca nandyāvartavad ekata eva sarvāḥ kiñcid vakrā yojyāḥ / . . . atra yuktiḥ / yantraikabhāge raso hy ārāmūlaṃ praviśati / anyabhāge tv ārāgraṃ dhāvati / tenākṛṣṭaṃ tat svayaṃ bhramatīti /*

29. Ibid., Golādhyāya, Yantrādhyāya, vv. 51cd-53ab:

    utkīrya nemim athavā parito madanena saṃlagnam //
    tadupari tāladalādyaṃ kṛtvā suṣire rasaṃ kṣipet tāvat /
    yāvad rasaikapārśve kṣiptajalaṃ nānyato yāti //
    pihitacchidraṃ tad ataś cakraṃ bhramati svayaṃ jalākṛṣṭam /

    This is followed by the description of another *perpetuum mobile* (vv. 53cd-56) in which the rim of the wheel is equipped with a series of pots which are successively filled with water from an overhead reservoir through a siphon, the full pots moving downwards and causing the wheel to turn. This cannot be really termed a perpetual wheel, as the reservoir has to be refilled. But this wheel is somewhat reminiscent of Su Sung's clock tower. Cf. S.R. Sarma, *Astronomical Instruments in Brahmagupta's Brāhmasphuṭasiddhānta,* op. cit., p. 74; see in this Volume, p. 63.

30. In his commentary on *Sūryasiddhānta* 13.22, p. 365; cf. S.R. Sarma, *Astronomical Instruments* . . ., op. cit., p. 74; see in this Volume, p. 63.

family of astrologers with connections at the Mughal court and may have seen or heard of automatic clocks and watches brought by Europeans as gifts to the Mughal court. He did not, however, realize that the first model of an automatically turning device was designed in India itself about a thousand years before him by the astronomer Brahmagupta.

This brings us to the end of the story of the *perpetuum mobile* and its Indian origin. In today's world of narrow loyalties, one is accustomed to ask to whom the credit should go: is it due to Brahmagupta for the origin of the ideas, or to the Islamic World for its elaboration and spread, or to the Occident for its practical application? Lynn White, quite rightly, sees these three kinds of endeavour as complementary to one another, when he says that

> it is an objective fact that, despite difficult communication, mankind in the Old World at least has long lived in a more unified realm of discourse than we have been prepared to admit.[31]

We may conclude by saying that when an anonymous painter in Europe depicts the Almighty God as an architect-engineer holding the globe of earth in one hand and a pair of compasses in the other,[32] or when Bhoja, King of Dhārā, conceives God as creating the universe in the form of a perpetual motion wheel,[33] they too seem to share a unified realm of vision.

31. L. White Jr., *Medieval Religion*, op. cit., p. 57.
32. The painting is preserved in the Nationalbibliothek, Vienna, and reproduced in J. Gimpel, op. cit., p. 145.
33. *Samarāṅgaṇasūtradhāra*, ed. T. Gaṇapatiśastrī, rev. V.S. Agrawala (GOS 25), Baroda, 1966, 31.1:

    bhrāmyad-dineśa-śaśi-maṇḍala-cakra-śastam
    etaj jagat-tritaya-yantram alakṣyamadhyam/
    bhūtāni bījam akhilāny api samprakalpya
    yaḥ santataṃ bhramayati smarajit sa vo 'vyāt//

    "He conceived the three worlds as a mechanical wheel,
    Its rim he fashioned from the orbits of the sun and the moon.
    The nave is hidden, but the creatures provide the motive power.
    Having crafted this world-machine,
    He lets it rotate for ever and ever.
    May that Śiva, the passion's conqueror, protect you."

4

Astronomical Instruments in Mughal Miniatures

0.1 The rich treasures of the Mughal miniature paintings have been studied quite fruitfully in recent years for reconstructing the material culture of the period.[1] Some of these paintings offer information also on astronomical instruments used in Mughal India in the sixteenth and seventeenth centuries. In a pioneering interpretation of a manuscript illustration from the *Hamzānāma,* Eric Forbes identified two Arab navigational instruments: a *kamāl* for measuring the stellar altitudes on the high seas and a portable mariner's compass with the magnetic needle suspended on a pivot.[2] In this paper, I shall discuss the astronomical instruments depicted in other miniatures. These miniatures can be grouped into two categories: while one group consists of miscellaneous scenes, the theme of the other group is the joyous celebration of the birth of a royal prince—be it Timūr, Akbar, or his sons Jahāngīr or Murād.

0.2 There are some seven miniatures, commissioned originally for illustrating the manuscript copies of the dynastic chronicles like

First published in: *Studien zur Indologie und Iranistik,* 16-17 (1992) 235-76.

1. Cf. Som Prakash Verma, *Art and Material Culture in the Paintings of Akbar's Court,* New Delhi 1978; Ahsan Jan Qaiser, *The Indian Response to European Technology and Culture (A.D. 1498-1707),* Delhi 1982; idem, *Building Construction in Mughal India: The Evidence from Painting,* Delhi 1988.
2. E.G. Forbes, "A 16th Century Indian Miniature illustrating two Arab Navigational Instruments" in: *Papers Presented [at the] International Conference on Science and Islamic Polity: Islamic Scientific Thought and Muslim Achievements in Science,* Islamabad 1983, vol. II, pp. 330-337.

the *Timūrnāma* and the *Akbarnāma*, which depict the birth of a prince in quite an elaborate manner. These are composite pictures consisting of several panels, each of which shows a different scene connected with the joyful event: the contented queen with the baby; maids attending hurriedly to various tasks; musicians and dancing girls performing with gay abandon; gifts being brought for the exalted new-born; distribution of alms to the poor; and, of course, astrologers casting the prince's horoscope (see Figs. 4.1-4.5).[3] The importance of the last-mentioned activity can be gauged from the prominence given in the composition to the astrologers' panel. Indeed, the compositions look as if they have two focii: the queen with the new-born prince on the one hand and the astrologers with their professional equipment on the other. The latter are shown measuring the birth time with a water clock or sand clock, determining the sun's altitude with a ring dial, finding the ascendant by consulting books of astronomical ṭables, and finally preparing the prince's horoscope.

0.3 Thus these miniatures represent more vividly what, for example, Abū 'l-Fazl records about Akbar's birth, in the typical courtier's style:

When the victory-seeking standards [i.e. Humāyūn] were leaving the fort of Amarkōṭ, Maulānā Chānd, the astrologer, who was possessed of great dexterity in the science of the astrolabe, in the scrutinizing of astronomical tables, on the construction of almanacs, and the interpretation of the stars,—was deputed to be in attendance at the portals of the cupola of chastity (Miryam Makānī, Akbar's mother) in order that he might ascertain

3. Besides the five miniatures reproduced here in Figs. 4.1-4.5, there are two more that depict the birth of a prince: *A Catalogue of the Indian Collection in the Museum of Fine Arts, Boston*, part VI: Mughal Paintings, by Ananda K. Coomaraswamy, Cambridge, Mass. 1930, Frontispiece and Pl. IV. The bottom part of the painting where Hindu and Muslim astrologers are seated is slightly damaged, but a ring dial held aloft by a Muslim astrologer can still be seen clearly. For a description of the painting, see *ibid.*, pp. 17-18. (ii) *Timūrnāma*, f. 284a, Khuda Bakhsh Library, Patna. Two astrologers are seated in a kiosk outside the palace gate. One is holding a book. Next to the other is a circular object suspended from a tripod. In the rotograph copy of the Ms. in Maulana Azad Library, Aligarh, it is not clear whether the circular object is a ring (= ring dial) or a disc (= astrolabe).

exactly the period of birth. He reported in writing to the exalted camp that, according to the altitudes taken by the Greek astrolabe, and by calculations based on the Gurgānī tables (Canons of Ulugh Beg), the figure of the nativity was as follows.[4]

What are the measurements Maulānā Chānd took? In Abū 'l-Fazl's words, Akbar's birth occurred when the altitude of Procyon was 38° and when 8 hrs. and 20 m. had passed from the beginning of the night of . . . Sunday 5th Rajab 949, lunar era, and [corresponding] to 6th Kārtik 1599, Hindū era, . . . 4 hrs. 22m. of the said night were remaining.[5]

That is to say, Akbar was born on the night between the 14th and 15th of October 1542 (Julian), 8 hours 20 minutes after the local sunset and 4 hours 22 minutes before the next local sunrise. These measurements of time and of the altitude of the star Procyon can easily be taken by means of an astrolabe.

One would, therefore, expect that in these miniatures the royal astrologers are shown holding the astrolabe, in accordance with Abū'l-Fazl's report. This is also the customary mode of depicting

4. *Akbarnāma*, I, p. 69. The "Greek" astrolabe is a pedantic allusion to the Greek origin of the instrument.
5. Ibid., I, pp. 53-54. Akbar's birth date is recorded in five different eras. I cite only the two relevant ones. Maheśa Ṭhakkura translated an abridged version of the *Akbarnāma* into Sanskrit, for which service he is said to have received the kingdom of Mithila from Akbar, cf. *Sarvadeśavṛttāntasaṃgraha or Akabaranāmā of Mahāmahopādhyāya Maheśa Ṭhakkura*, ed. Subhadra Jha, Patna 1962, pp. xiii-xvi. If so, it is quite a heavy price, when one considers the quality of translation. The passage dealing with Akbar's birth is rendered thus (ibid., p. 17):

   asmin samaye viṃśatighaṭikāsu viṃśatipalayutāsu vyatītāsu . . . ekona-pañcāśad-adhika-śatanavake 'tīte cāndramānasya pañcamarātrau ravivāsare, navanavaty-adhika-pañcadaśaśateśv atīteṣu vikramārka-rājyāt kārtike māsi ṣaṣṭyāṃ tithau . . . dvāviṃśatipalādhika-daśaghaṭi-kāvaśeṣāyāṃ rajanyām . . .

   Actually 8 h. 20 m. are equal to 20 *ghaṭikās* and 50 *palas* (and not 20 *gh.* 20 *p.* as the Sanskrit reads); 4 h. 22 m. are equal to 10 *gh.* 55 *p.* (not 10 *gh.* 22 *p.*). Maheśa, therefore, converted only the hours into *ghaṭikās*, but in the *palas'* place just put the same number as that of minutes!

the astrologers/astronomers elsewhere in the Islamic World.[6] But the Mughal artist seems to have other notions, and one rarely sees the astrolabe in these miniatures. This is surprising on several counts.

1. Astrolabe (Yantrarāja)

1.1 In Islamic culture, the astrolabe[7] enjoyed a high reputation as the "jewel of mathematics". It is a highly versatile instrument, capable of several functions. While the ring dial can be used only in the daytime for measuring the sun's altitude and time, with the help of the astrolabe time can be measured both in the daytime and at night, in equal hours or in unequal hours. One can take the altitudes, and read directly from the dial the ascendant (*lagna*), i.e. the point on the ecliptic which is just above the horizon at the desired moment. In horoscopy, the ascendant is of great importance, because it is from this point that the zodiac circle is divided into 12 houses, and the planetary configuration in each house has a bearing on a specific aspect of the child's life. The astrolabe enables one to read the configurations without having to make long and tedius computations.

1.2 The astrolabe was introduced into India apparently during the reign of Fīrūz Shāh Tughluq in the second half of the fourteenth century. His contemporary biographer Shams-i-Sirāj 'Afīf reports

6. Cf. e.g., miniatures illustrating the "Birth of Child" and the "Casting a Horoscope" from the Ms. of al-Harizi's *Maqāmat*, Bibliothèque Nationale, Paris, reproduced in: René Taton, *Ancient and Medieval Science*, tr. into English by A.J. Pomerans, London 1963, Pls. 33 and 34 between pp. 416-417. The former is reproduced also in Turner, I.1, p. 27. See also the oft-reproduced miniature "Astronomers at Work" from the Ms. of *Kitāb Rasda al Munajjimīn*, of AH 813 (AD 1410/11), Istanbul University Library, F-1418, Pl. 1b.
7. On the astrolabe, see Robert T. Gunther, *The Astrolabes of the World*, Oxford 1932; Willy Hartner, "The Principle and Use of the Astrolabe", and "Asṭurlāb", reprinted in: *Oriens-Occidens*, [vol. I], Hildesheim 1968, pp. 287-319; Turner, I.1; David A. King, "Astronomical Instrumentation in the Medieval Near East," reprinted in his *Islamic Astronomical Instruments*, London 1987, article no. I.

that Fīrūz always had a *nisfī* astrolabe by his side.[8] Fīrūz also had an astrolabe painted on his banners. One such banner was hung on the Aśoka pillar, which he got transported from the Shiwalik hills and set up on a specially constructed building in his citadel at Delhi.[9]

Of the Mughal emperors, Humāyūn (reign 1530-56) is said to have "extraordinary excellence in the astrolabe, globe and other instruments of the observatory."[10] His sister reports that he "took the astrolabe into his blessed hand" and himself chose the propitious moment for his marriage.[11] More important is the fact that under his active patronage, astrolabe manufacture started at Lahore. Specimens of astrolabes produced by several generations of a single family are extant today. The earliest member of this astrolabist family is Ustād Shaikh Allāhdād Asṭurlābī Humāyūnī Lāhūrī. From this name, Sulaiman Nadvi conjectured that Humāyūn invented a special type of astrolabe called Humāyūnī asṭurlāb and therefore the family of its manufacturers bore the title Asṭurlābī Humāyūnī.[12] The extant astrolabes produced by this family, however, do not bear out Nadvi's contention; they are modelled quite faithfully after the contemporary Persian instruments.[13] But there is no doubt that Humāyūn promoted the manufacture of astrolabes, and also of celestial globes. We have also seen that Maulānā Chānd was an expert in the use of the astrolabe.

8. I.e. a bipartite astrolabe having 45 almucantar lines, each one representing 2° of altitude, cf. Gunther, op. cit., I, pp. 7-8.
9. Syed Athar Abbas Rizvi, *Tughluq-kālīna Bhārata*, part 2, Aligarh 1957, p. 146 (Hindi tr. of 'Afif's *Tārīkh-i Fīrūz Shāhī*); see also S.A.K. Ghori, "Scientific Exchanges between Soviet Central Asia and India during Medieval Times", in: B.V. Subbarayappa (ed.), *Indo-Soviet Seminar on Scientific and Technological Exchanges between India and Soviet Central Asia (Medieval Period)*, New Delhi 1985, pp. 78-89, esp. 87-88
10. Syed Suleiman Nadvi, "Some Indian Astrolabe-Makers", *Islamic Culture*, 9 (1935), pp. 621-631, esp. 622-623.
11. *The History of Humāyūn (Humāyūn-Nāma) by Gul-Badan Begum*, tr. A.S. Beveridge, Delhi 1972, p. 51.
12. Nadvi, op. cit., p. 626
13. Turner, I.1, p. 26.

1.3 Hindu and Jaina astronomers were no less enthusiastic about the advantages of the astrolabe, and wrote a number of manuals on its construction and use. The first such work, entitled *Yantrarāja*, was written by the Jaina monk Mahendra Sūri in 1370. In this book, Mahendra Sūri calls the astrolabe the "king of instruments" (*yantrarāja*).[14] His pupil Malayendu Sūri wrote a commentary on this work in about 1382, where he informs that Mahendra was associated with Fīrūz Shāh Tughluq's court. Apparently Fīrūz was not only interested in using the astrolabe himself but he also encouraged the Jaina Sūri to compose a work on it in Sanskrit.

Within a quarter century after the Sūri's work, i.e. about 1400, Padmanābha wrote the *Yantracintāmaṇi* (also known as *Yantrakiraṇāvalī*) and devoted the first chapter to the astrolabe.[15] Then in 1428, Rāmacandra Vājapeyin, a resident of Naimiṣāraṇya (near modern Lucknow) discussed the astrolabe quite extensively in his *Yantraprakāśa*.[16] This unique text describes the construction and use of some 35 astronomical instruments—perhaps the largest number ever dealt with in a Sanskrit work. Some of these are traditional Indian instruments, or their variants, and some are clearly of Islamic origin. As we shall see in the following pages, the value of the *Yantraprakāśa* cannot be underestimated in tracing the history of individual instruments. The major part of this work is devoted to

14. *The Yantrarāja of Mahendra Sūri* with the commentary by Malayendu Sūri, and *the Yantraśiromaṇi of Daivajñacūḍāmaṇi Viśrāma*, ed. K.S. Raikva, Bombay 1936. David Pingree, *Jyotiḥśāstra. Astral and Mathematical Literature*, Wiesbaden 1981, p. 53, observes that "After Mahendra a number of other texts describing traditional Indian instruments were composed in Sanskrit, almost all of them in Gujarat and Rajasthan, but the astrolabe was generally neglected." In fact, the astrolabe received the greatest attention, even outside Gujarat-Rajasthan region, as the following lines will show.
15. On this and the following texts, see Yukio Ohashi's comprehensive survey, "Sanskrit Texts on Astronomical Instruments in the Delhi Sultanate and Mughal Periods," *Studies in History of Medicine and Science*, 10-11 (1986-87), pp. 165-181.
16. Ms. no. 975 of 1886/92 of the Bhandarkar Oriental Research Institute, Poona (BORI); and Ms. no. G-1363 of the Asiatic Society, Calcutta (ASB). See also § 3.1 and § 5.5 below.

the astrolabe, which is called here *sulabhā*, another significant name meaning that with this instrument several types of measurements become easy.

In the reception of scientific ideas from Islamic culture and in their dissemination in India, Jaina monks seem to have played a prominent role. About the end of the fifteenth century, another Jaina monk, Muni Megharatna, pupil of Vinayasundara of Vaṭagaccha, wrote the *Usturalāvayantra* in 38 stanzas, replete with Arabic-Persian technical terms.[17]

Two seventeenth century texts discuss the astrolabe along with other instruments. Thus Viśrāma devotes the third chapter of his lucidly written *Yantraśiromaṇi* (1615) to the astrolabe.[18] It was also discussed in Nityānanda's *Siddhāntarāja*, which was completed in 1639.[19]

In the first half of the eighteenth century, astronomers at the court of Sawai Jai Singh (1688-1743) translated several books on astronomical instruments from Arabic or Persian into Sanskrit, and also wrote some independent works. Four of these deal with the king of instruments. The *Yantrarājaracanā*, attributed to Jai Singh himself, deals exclusively with the astrolabe.[20] The eighth chapter of a compilation entitled *Yantraprakāra* also explains the use of this instrument.[21] Furthermore, Naṣīr al-Dīn al-Ṭūsī's *Risālah-i Bist Bāb dar Ma'rifat-i Usṭurlāb* was rendered into Sanskrit under the

17. Cf. Ambalal P. Shah, *Jaina Sāhitya kā Bṛhad Itihāsa*, vol. V, Varanasi 1969, p. 180; Agar Chand Nahata, "Usturlāva Yantra saṃbhandhī eka Mahatvapūrṇa Jainagrantha," *Jaina-Siddhānta-Bhāskara,* XVIII.ii, pp. 119-128. Shri Hazari Mull Banthia (Kanpur) very kindly obtained a xerocopy of this article for me. The only Ms of the *Usturalāvayantra,* with a Sanskrit commentary and Rājasthānī gloss, is in the Anup Sanskrit Library, Bikaner.
18. See n. 14 above.
19. See Ohashi, op. cit. (n. 15).
20. *The Yantrarājaracanā of Jayasiṃhadeva*, ed. Kedāranātha Jyotirvid, Jaipur 1953, to which is appended, on pp. 17-19, the *Yantraprabhā* of Srīnātha.
21. *Yantraprakāra of Sawai Jai Singh*, ed. & tr. S.R. Sarma, Supplement to *Studies in History of Medicine and Science*, 10-11 (1986-87), pp. 20-21, 61-63.

title *Yantrarājavicāraviṃśādhyāyī.*[22] Jai Singh also caused the composition of the *Sarvadeśīya-jarakālīyantra* on the universal plane astrolabe invented by the Moorish astronomer al-Zarqālluh in Spain at the end of the eleventh century.[23]

Three more works were composed after Jai Singh's time. Śrīnātha Chagāṇī wrote the *Yantraprabhā,*[24] which is an abridged version in verse of Jai Singh's *Yantrarājaracanā.* In 1772 Nandarāma composed the *Yantrasāra* in which the astrolabe was discussed along with a number of other instruments.[25] Finally, Mathurānātha Śukla wrote his *Yantrarājaghaṭanā* in 1782.[26]

Manufacture of the "Hindu astrolabes", i.e. astrolabes with markings in Devanāgarī script and with the time scale divided into *ghaṭīs* instead of hours received encouragement through Jai Singh's astronomical activities, but earlier specimens are also known.

1.4 In spite of its popularity among the Muslim scholars and in spite of its enthusiastic adoption by Hindu and Jaina astronomers, none of the nativity paintings depict the astrolabe. In the other group, an astrolabe can be seen in two miniatures only. In the "Noah's Ark" (Fig. 4.11), the ship's pilot is seated on the poop deck on the left-hand side of the picture and is taking the altitude by means of an astrolabe. In front of him, there is an open book on a stand, presumably a book of astronomical tables. Again, in a miniature depicting a venerable astronomer in idyllic surroundings

22. *Yantrarājavicāraviṃśādhyāyī of Nayanasukhopādhyāya,* ed. Vibhuti Bhushan Bhattacharya, Varanasi 1979. The grounds for the attribution of this translation to Nayanasukha are quite doubtful. At the end of the Ms. on which this edition is based, there is an addendum by a later hand: *iti Nayanasukhopādhyāyakṛta-Yantrarājavicāraviṃśādhyāyī Arabītāḥ Samskṛte nītā* (see the facsimile reproduction). No other Ms. has this colophon, nor was the work translated from the Arabic (the original is in Persian).
23. I have seen Ms. no. 5483 of the Khas Mohor Collection of the Maharaja Sawai Mansingh II Museum, Jaipur. An identical text is inserted into the Spaṣṭādhikāra in: *Siddhāntasamrāṭ Jagannāthasamrāḍ-viracitaḥ,* ed. Muralidhara Caturveda, Sagar 1976, pp. 96-105.
24. See n. 20 above.
25. I have used Ms. 504 of 1892/95 from BORI.
26. Cf. Ohashi, op. cit. (n. 15).

(Fig. 4.9), a small astrolabe can be seen on the mat in front of him. But in both cases, the astrolabe is pooriy drawn, it is just the outline without any details.

A.J. Turner thinks that "astrolabes were far from being widely spread in Mughal society, and perhaps remained exclusive preserve of a court coterie."[27] In view of the large number of Sanskrit works written on this instrument, it is not possible to agree with Turner. Yet it is difficult to explain the Mughal painter's reluctance to draw this instrument, except to say that perhaps the ring dial was more widely used.

2. Water Clock (Ghaṭikāyantra)

2.1 Before taking up the ring dial which our miniature painters prefer to the astrolabe, it will be convenient to deal with two types of time-measuring instruments in these miniatures. In two nativity paintings (Figs. 4.1, 4.5), the astrologers use the sand clock for determining the birth time. In two others (Figs. 4.2, 4.3), there are finely drawn water clocks of the sinking bowl variety.[28] While the sand clock in the Mughal miniatures has been noticed and commented upon, the water clock escaped the attention of art historians until now.

This type of water clock consists of a hemispherical bowl, with a fine aperture at the exact centre of its bottom. When this bowl is placed on the surface of the water in a larger vessel or basin, water enters the bowl through the hole, fills the bowl gradually and causes it to sink to the bottom of the basin. The hole is so made that the bowl fills up and sinks in a specific interval of time, usually 24 minutes, called *ghaṭī* or *ghaṭikā* in Sanskrit.

In Fig. 4.2, which depicts Akbar's birth, two astrologers are seated in front of the emperor and are explaining to him the infant's horoscope, which is presumably written on the sheet(s) of paper lying on the carpet between the two. A ring dial is also on the carpet, to the right of the horoscope. A step below, just in front of

27. Turner, I.1, p. 26.
28. On water clocks in general, see Turner, I.3.

the horoscope, one can see the water clock; the floating bowl and the larger vessel are carefully executed.

The subject of Fig. 4.3 is the birth of Salīm, the future Jahāngīr. Four astrologers are seated on a carpet outside the entrance to the harem. In front of the carpet can be seen the water clock, with a highly ornamental basin. Both the bowl and the basin appear to be gilded or golden, which is but appropriate in the instrument for measuring the birth time of the heir to the Mughal throne.[29]

2.2 I have discussed elsewhere[30] the history of this type of water clock, spanning some 1600 years from the fourth century to the nineteenth. Known as *Ghaṭikāyantra* or *Ghaṭīyantra*, it was described by almost all astronomers since Āryabhaṭa. In the seventh century, it wholly replaced an older type, viz., the cylindrical outflow clock called *Nāḍikāyantra*. In royal palaces, Buddhist monasteries and other public places, time was regularly measured by means of the *Ghaṭikāyantra* and broadcast at certain intervals by beating drums and blowing conch-shells. Public and private endowments were made for the maintenance of this institution. The same situation prevailed in north-western India in al-Bīrūnī's time.

Subsequently, but before the fourteenth century, the drum and conch-shell were replaced by the gong to announce the passage of time intervals. Fīrūz Shāh Tughluq was so impressed by this method of measuring and announcing time that he set up a water clock and gong (*tās-i ghariāla*) on the entrance gate of his palace. He also got this device depicted on some of his gold coins, which he used as gifts to persons of high rank.[31] Fīrūz's enthusiasm implies that no such instrument was known to the Islamic world.

This conclusion is reinforced by Bābur's (reign 1526-30) detailed description of this device in his memoirs. He readily adopted the instrument and also the Indian division of time into *praharas, ghaṭīs* and *palas*, and introduced an innovation in the method of striking

29. Writing in 1615, Viśrāma in fact recommends that the whole apparatus be gilded (*hemāṃbuliptaṃ sakalaṃ vidheyam, Yantraśiromaṇi* 2.3, cf. n. 14).
30. "Water Clocks and Time Measurement in India", to appear in *Aligarh Journal of Oriental Studies* 6 (1989).
31. Syed Athar Abbas Rizvi, op. cit., pp. 108-109.

the "hours".[32] Thus the *Ghaṭikāyantra* was in regular use at the Mughal court, and not just on special occasions. The two miniatures just mentioned are unique in that they contain the only pictorial representation of this water clock.

3. Sand Clock (Kācayantra)

3.1 In other paintings, time is measured with a sand clock instead of the water clock. The appearance of the sand clock in several Mughal miniatures from the end of the sixteenth century led historians to conclude that it was imported from Europe during Akbar's reign.[33]

However, the sand clock was known in India much before these miniatures were painted. It was described for the first time in 1428 by Rāmacandra Vājapeyin in his *Yantraprakāśa*. He calls it *Kācayantra*, "glass instrument" , and explains that it measures one *ghaṭī*. I give below his description of the *Kācayantra*, along with his own commentary on it:

kācasya pātryau vadane yute sṛjet
ūrdhvām bhṛtāṃ vālukayāvadhārayet /
ghaṭyāṃ tu tatpātata eva vālukā-
dharordhvagā śvetatarā punas tathā //

ghaṭyāṃ vālukāmokṣayogyāsya-samapramāṇa-kācapātryor ekāṃ vālukāpūrṇāṃ kṛtvā tayor mukhaṃ yutaṃ baddhvā bhṛtopari dhāryā / ghaṭyāṃ sikatāpāte upariṣṭhā śvetādhaḥsthā kṛṣṇā / punar vinimayatvāt sthāpane ghaṭyante tathā /[34]

32. *Bābur-Nāma* (*Memoirs of Bābur*), tr. Annette Susannah Beveridge, reprint: Delhi 1979, I, pp. 516-517. Strangely enough, Abū 'l-Fazl's description of the water clock is rather inaccurate and this is further aggravated by Jarrett's translation and Sarkar's annotation (*Ā'īn-i Akbarī*, III, pp. 17-18, n. 15a). Abū 'l-Fazl says that the bowl measures "twelve fingers in height and breadth". If a hemispherical bowl measures 12 fingers in breadth (i.e. the diameter at the mouth), its height will be 6 fingers only. These measurements are given in almost every Sanskrit source, see my paper mentioned in n. 30.
33. S.P. Verma, *Art and Material Culture . . .*, p. 112; Qaisar, pp. 36-37.
34. BORI 975 of 1886/92, f. 70r; ASB G-1363, f. 90v. Orthography silently emended.

"Make two glass ampoules *(pātrī)* joined mouth to mouth. The upper one should be filled with sand. Owing to its trickling down, after one *ghaṭī* [all] the sand will be below, and the upper [ampoule becomes] white. [Do] the same again."

"Of two glass ampoules of equal size, having mouths capable of discharging [a certain amount of] sand in one *ghaṭī*, fill one with sand, join the mouths and tie the juncture securely, and then set up [the apparatus] so that the full ampoule is above. When [all] the sand trickles down [into the lower ampoule] within one *ghaṭī*, the upper [ampoule becomes] white (i.e. empty and transparent) and the lower one dark (as it is full of sand). Then exchange their places (i.e. put the clock upside down); after the lapse of one *ghaṭī*, the same [thing happens once again]."

This is an unambiguous description of the sand clock. It may be noted that Rāmacandra's clock measures the Indian time unit of *ghaṭī*, while the European sand clocks were made to measure the hour, its multiples or fractions. If the sand clock was transmitted from Europe to India, where it was adapted to the Indian system of reckoning time in *ghaṭīs*, such a transmission ought to have taken place quite some time before 1428, at the latest by 1400. Even today adapting an imported technology to suit local conditions requires a long time.

3.2 In Europe itself the first indisputable mention of the sand clock occurs in the English naval records of 1345/46 and its pictorial representation, in a fresco at Sienna in Italy, pertains to about the same period. Of course, in the fifteenth century, the sand clock became a common object of daily life. It was employed in measuring the length of the lesson at school and university, and more particularly, the length of the sermon at church. Also, in painting and sculpture, it became the symbol of Father Time. [35]

Now, if Europe had been the source of the sand clock which was described by Rāmacandra in 1428, the only possible means of its

35. Lynn White, *Medieval Technology and Social Change*, Oxford 1962, pp. 103, 165-166; idem, *Medieval Religion and Technology*, Berkeley 1978, pp. 193, 220, Fig. 5; Turner, I.3, pp. 75-84; R.T. Balmer, "The Operation of Sand Clocks and Their Medieval Development," *Technology and Culture*, 19 (1978), pp. 615-632.

transmission, say between 1350 and 1400, should be through the Islamic World, but there is no trace of such mediation having taken place. The Islamic World itself did not know the sand clock until the late sixteenth century when it was imported from Europe: in the well-known painting of Taqī al-Dīn's observatory (1577/78) at Istanbul, one can see two sand clocks of different sizes.[36]

Therefore, Rāmacandra's description of the sand clock is based neither on a European source, nor on an Islamic prototype. Furthermore, Jñānarāja in his *Sidhāntasundara* describes both the sinking bowl type of water clock (*ghaṭī*) and the sand clock (*kācayantra*) in two halves of a single verse as though the two instruments were equally popular at his time.[37] His book is dated 1503, i.e. almost a century before the European travellers could have brought sand clocks to India as gifts or as articles of sale. Hence, we must conclude, at least tentatively, that the sand clock developed independently both in India and in Europe.

3.3 This is not to suggest that all the sand clocks depicted in the Mughal miniatures are of Indian origin. At the end of the sixteenth century, Europeans brought to India sand clocks that measured the hour or its fractions,[38] and these were accepted at the Mughal court. Although the Mughals had adopted the traditional Indian division of time into *praharas, ghaṭīs* and *palas* they did not give up the hours and minutes which prevailed in the Islamic World. Probably

36. Reproduced often, e.g. in Syed Hossein Nasr, *Islamic Science, An Illustrated Study*, London 1976, Pl. 65, p. 113; Turner, I.1, p. 25.

37. *Siddhāntasundara* 17.30:

ghaṭadalaghaṭitā ghaṭī niruktā
talasuṣirā palaṣaṣṭhitaḥ kapūrṇā /
murajasamaṃ athācchakācayantraṃ
talasuṣiraṃ kṣaradalaśarkarāḍhyam //

"The *Ghaṭī[-yantra]* is defined as the one which is made of a hemispherical bowl with a hole at the bottom and which fills with water (*ka*) in 60 *palas* (= 1 *ghaṭī*). Next, the transparent *Kācayantra* is shaped like *muraja*-drum with a hole at the bottom (i.e. mouth of the ampoule) and is filled with sand which trickles down." The text is from India Office Library Ms. no. 2002. A transcript was kindly sent by Mr Yukio Ohashi (Tokyo).

38. Qaisar, pp. 36-37, 76-77.

they reckoned time in *ghaṭīs* and *palas* for the daily routine as the local people did, but used hours and minutes for scientific or scholarly purposes. The astrolabes made at Lahore had the time scale divided in hours. It may also be recalled that Abū 'l-Fazl recorded the time of Akbar's birth in hours and minutes whereas Maheśa Ṭhakkura converts them into *ghaṭīs* and *palas* in his Sanskrit rendering of the *Akbarnāma.*

Therefore, the Mughal court may have used the European sand clocks side by side with the indigenous ones to measure hours with the one and *ghaṭīs* with the other. *Ghaṭīs* are mentioned in the *Ā'īn-i Akbarī* quite often, e.g. in connection with the game of polo, it is stated that after each *ghaṭī,* two players retired and their place was taken by two other fresh players.[39] These *ghaṭīs* may have been measured by the water clock or with the indigenous sand clock.[40]

In 1668-72, John Marshall noticed such indigenous sand clocks. He says that "In some places, as at Patna, they have glasses with sand in them, made like our houre-glasses in England, which are exact gurry (*ghaṭī*)."[41] Here Marshall is making a clear distinction between the English sand clock that measured the hour and the Indian device that measured the *ghaṭī.*

But it is difficult to say whether the Mughal court began using the sand clock only after the advent of the Europeans or even before that. In any case, the European sand clocks seem to have become quite popular. The Persian word for the sand clock *shīsh-i sāat*, literally "hour glass", shows that the designation was clearly inspired by the European specimens. It is under this name that the device was described for the first time in a dictionary by Faizī Sarhindī in 1598.[42]

39. *Ā'īn-i Akbarī*, I, p. 309.
40. The advantage of the sand clock is that it is easily portable, say to the polo field. But it cannot be graduated to read fractions of its duration. Instead, one has to use a set of two, four, six, or eight sand clocks of different durations, but such variants do not seem to have been in vogue in India. The water clock, on the other hand, is a bit cumbersome to carry, but it is possible to graduate the bowl empirically.
41. As cited in Qaisar, p. 76.
42. Irfan Habib, "Cartography in Mughal India", *Medieval India, A Miscellany*, 4 (1977), pp. 122-134, esp. 132.

3.4 Coming to the depiction of the sand clock in the Mughal miniatures, it occurs in six paintings (Figs. 4.1, 4.5, 4.6, 4.8, 4.9 and 4.10) but its iconography varies from painting to painting. The European sand clocks are invariably housed in a protective and supporting frame made of metal or wood. Moreover, until the beginning of the eighteenth century, i.e. during the period with which we are concerned here, sand clocks were manufactured in Europe in the following manner. The two ampoules were blown separately, then a perforated diaphram of brass was inserted between the mouths of the two ampoules and the joint was "secured by pitch, wax or putty bound over with canvas or another fabric and lashed tight with crisscrossed thread or gilt wire."[43] Thus in the iconography of the European sand clock, there are three main elements: the frame, the triangular ampoules, and the bulging joint crisscrossed by lines.

Since we do not know how the Indian sand clock was manufactured except what Rāmacandra informs us, it is not certain if the absence of any of these three iconographic elements should point to an indigenous make. In our miniatures, there are cases where one or two elements are missing. Thus in two miniatures (Figs. 4.5, 4.10), the clock consists of just two ampoules without a frame.[44] In Fig. 4.1, there is a frame, but the whole picture is highly stylised, the sand clock in a frame being represented by just a rectangle with its two diagonals. In Fig. 4.9, the sand clock is faintly visible—to the proper left of the astronomer, at the back wall of the hut, upon a small table and serving as a stand for an open book—but there is no doubt that the frame is of European make.

In Fig. 4.6, the highly embellished sand clock has an allegorical role. Jahāngīr is seated on a giant sand clock and is naturally conscious of the "falling sands of time". That is why he prefers the company of the Ṣūfī saint to that of the kings, as the caption of the painting declares. The gilt frame and the middle crisscrossed by

43. Turner, I.3, pp. 75-76.

44. This may not prove anything. Of the sand clocks drawn in the painting of Taqī al-Dīn's Istanbul observatory, the larger one is with a frame but the smaller one is without.

gold thread suggests that the picture was drawn after an ornate European sand clock that was received in gift by the emperor.[45]

Finally, the portrait of the astrologer with all his paraphernalia (Fig. 4.8) has a finely drawn sand clock, but it has two remarkable features. In this monochrome marginal painting, only sand was given a different colour—sky blue. This may have been inspired by the European specimens, for the filling was rarely real sand in Europe. One experimented with various substances and these may also have been coloured. The second feature is that the two ampoules appear to have been blown as one single piece, because one can see the continuous stream of the falling sand. But this technique of manufacturing sand clocks commenced in Europe only from 1760-1770,[46] whereas our picture was painted in Jahāngīr's atelier between 1605 and 1627. Does this mean that this new technique was available in India a century earlier than in Europe, or is this a case of artistic licence? Until actual specimens can be examined, it is safe not to draw too many conclusions from a single painting.

To conclude this account of the sand clock: though it was known in India since 1428, its pictorial representation in the Mughal miniatures appears to have been largely influenced by the specimens imported from Europe.

4. Celestial Globe (Bhagolayantra)

4.1 A globe is depicted in two miniatures. In Fig. 4.9, the astronomer in his country idyll is surrounded by various astronomical instruments, including a globe on his proper right. In Fig. 4.7, there are two soaring angels, one holding a globe and the other a ring dial (on the interpretation of this picture, see § 5.2 below). In both paintings, the globe is a solid spherical mass, without any markings or supporting stand, but there is no doubt that it is supposed to

45. Stuart Cary Welch, *Imperial Mughal Painting*, London 1978, p. 82: "Like the hour-glass throne, which may have been based on a small gilt-bronze and glass original, the idea of allegorical state portraits came from Europe, as did the cupids. . . ."
46. Turner, I.3, pp. 75-76.

represent the celestial globe, known to the Islamic World as *al-Kura* or *al-Kursī*.[47]

4.2 Described originally by Ptolemy in his *Almagest* (Book VIII, Chapter 3) for plotting and marking the star positions, it was adopted and improved upon both in medieval Europe and in the Islamic World. European portraits of the astronomer show him usually with a pair of compasses plotting the star positions on the celestial globe, e.g. in Jost Amman's illustration to Hans Sachs' doggerel on the "Astronomus".[48] During the Renaissance, automatically rotating celestial globes with clockwork mechanism were made in great numbers.[49] In the Islamic World also various treatises were written on this instrument and innovations were made in its construction.[50]

4.3 Of all the instruments under discussion here, the celestial globe is the one instrument that was introduced into India by Mughals, more specifically by Humāyūn whose expertise in its use was already mentioned above. Under his patronage, the Lahore family of astrolabists produced some celestial globes also, and the extant specimens were surveyed by Sulaiman Nadvi in 1935.[51] These were made of bronze, and the star positions were marked by inlaid silver points.

As compared to the astrolabe, it had a limited reception among the Hindu astronomers. The only mention that I know of is by Nṛsiṃha who gives a fairly long description in his commentary (written in 1621) on Bhāskara's *Siddhāntaśiromaṇi*.[52] Likewise,

47. King, op. cit., p. 4 (cf. n. 7 above).
48. *Jost Ammans Stände und Handwerker mit Versen von Hans Sachs,* Frankfurt a.M. 1568, reproduced in: Ludolf von Mackensen, *Die erste Sternwarte Europas mit ihren Instrumenten und Uhren; 400 Jahre Jost Bürgi in Kassel*, München 1982, 2nd edn., p. 21.
49. Ibid., passim.
50. R. Lorch, "al-Khāzin's Sphere that Rotates by itself", *Journal for the History of Arabic Science,* 4 (1980), pp. 287-329.
51. "Some Indian Astrolabe-Makers," *Islamic Culture*, 9 (1935), pp. 621-631.
52. *Siddhānta-Śiromaṇi of Bhāskarācārya* with his autocommentary *Vāsanābhāṣya* and *Vārttika* of Nṛsiṃha Daivajña, ed. Murali Dhar Chaturvedi, Varanasi 1981, p. 438.

there do not seem to be many "Hindu" globes, i.e. with the markings in Devanāgarī script. Sawai Jai Singh got one made which is now deposited in the storerooms of his Jaipur observatory. Another was once in the personal collection of David Eugene Smith and was supposed to have been made ca. 1600.[53]

5. Ring Dial (Cūḍāyantra)

5.1 Now we come to the observational instrument so consistently depicted in most of our miniatures. In six of the seven known paintings of the nativity (Figs. 4.1-4.5),[54] the ring dial is clearly drawn. In Fig. 4.3, the thickness of the ring and the inner concave surface can be clearly seen.

In the other group of miniatures, there is a portrait of the astrologer with all his equipment arrayed around him (Fig. 4.8). With his left hand, he is holding aloft a ring dial, the markings on which are clearly visible. This painting belongs to a series of portraits of different professions with their tools of trade accurately drawn. It is significant that this professional portrait of the astrologer, like the nativity paintings, should depict him holding the ring dial in a characteristic posture, and not with the astrolabe as was done elsewhere in the Islamic World. In Mughal India then, at least in popular conception, the ring dial was the emblem par excellence of the astrologer/astronomer.

5.2 In a miniature painted for Shāh Jahān (1628-1658), the simple ring dial is elevated to become the symbol of Time itself. Shāh Jahān commissioned a series of portraits of his forefathers, which contain highly interesting assemblage of marginal pictures. On the top margin of each portrait, there is a pair of angels holding various objects. In the portrait of Humāyūn, with which we are concerned

53. For a picture, see his *History of Mathematics*, II, Boston 1925, p. 365. After writing the above, I had occasion to consult Emilie Savage-Smith, *Islamicate Celestial Globes, Their History, Construction, and Use*, Washington, D.C., 1985, where this globe is described (p. 245) and reproduced (Fig. 25). She attributes it to the workshop of Bahlūmal Lāhūrī (mid-19th century). The globe is now at Columbia University, New York.

54. On the other two miniatures, see n. 3 above.

here, he is shown seated in a landscape.[55] On the top margin, two angels are holding a crown over his head, suggesting his universal sovereignty. In addition to the crown, one angel is holding a globe and the other a ring dial, perhaps as symbols of cosmic Space and Time (Fig. 4.7).

The motif of soaring angels holding a crown above the head of a personage is no doubt borrowed from the European paintings of the Madonna which were popular at the Mughal court.[56] But the ring dial did not belong to the European motif, it was added by the Indian artist. Thus in the conception of the Mughal artist—for at least three generations from Akbar to Shāh Jahān—the ring dial was inseparable from the astronomer.

5.3 It may have been so in real life as well, because it is a handy tool, though not very accurate, for measuring the sun's altitude. A small hole in the breadth of the ring allows the sunlight to pass through and fall upon the inner concave surface on the opposite side, which is graduated in degrees to measure the altitude. Local time can also be measured directly if the inner surface is provided with separate scales for each month, and each scale is divided into so may *ghaṭīs* as there are in a half day (from the sunrise to the noon) in that month.

5.4 But what are the antecedents? Because of its frequent occurrence in paintings depicting the Mughal court, one is apt to think that it may have been of Islamic origin. But according to competent authorities on Islamic astronomical instruments whom I consulted, the ring dial was not known to the Islamic World.[57] On the other hand, the Sanskrit astronomical texts show that it has almost as long a history in India as the *Ghaṭikāyantra*.

It was first described by Āryabhaṭa around the beginning of the sixth century in his *Āryabhaṭasiddhānta*. He calls it *Cakrayantra* and describes two holes in the breadth at diametrically opposite

55. Cf. Constance A. Bond, "A Priceless Collection Rediscovered," *SPAN*, 30.2 (May 1989), pp. 2-5 and the cover page.
56. Cf. Qaisar, pp. 88-94.
57. Professors E.S. Kennedy (Princeton) and David A. King (Frankfurt).

points.[58] The instrument is described by Varāhamihira also, in the middle of the sixth century, in his *Pañcasiddhāntikā*. He does not give it a name, but provides only a single hole and intends it to be used for measuring the sun's meridian zenith distance:[59]

> Take a circular hoop, on whose circumference the 360 degrees are marked, whose diameter is equal to one hasta, and which is half an aṅguli broad. In the middle of the breadth of that hoop make a hole. Through this small hole made in the circumference allow a ray of the sun at noon to enter in an oblique direction. The degrees, intervening at the lower half of the circle between (the spot illumined by the ray and) the spot reached by a string hanging perpendicularly from the centre of the circle, represent the degrees of the zenith-distance of the midday sun.[60]

Of course, a string cannot be suspended from the "centre" of a ring, but a string let down from the topmost point of the ring will pass through the notional centre and represent the vertical.

5.5 In the *Yantraprakāśa,*[61] Rāmacandra describes three varieties of this instrument, viz., *Valaya* (literally "circular hoop"), *Cūḍā*

58. The *Āryabhaṭasiddhānta* is not longer extant, but its chapter on instruments survives in quotations, notably in Rāmakṛṣṇa Ārādhya's commentary (1327) on the *Sūryasiddhānta*. K.S. Shukla gathered the stanzas of this chapter and interpreted in: "Āryabhaṭa I's Astronomy with Midnight Day-reckoning," *Gaṇita* 18 (1967), pp. 83-105. Āryabhaṭa describes the *Cakrayantra* thus (ibid., p. 93):

    bhagaṇāṃśāṅkitaṃ cakraṃ sarandhraṃ viṣuvaty atha //10//
    dhanū ravyunmukhaṃ kṛtvā cāpavac cakrayantrakam /
    kalpayel lambaśaṅkor vā chāyānāḍyaś ca yaṣṭivat //11//

59. *Pañcasiddhāntikā* 14.21-22:

    samabhagaṇāṅkakacakram ardhāṅgulabahalam āyataṃ hastam /
    vistāramadhyabhāge chidraṃ tadgāmi tiryak ca //
    madhyāhnārkamayūkhaṃ praveśya sūkṣmeṇa paridhivivareṇa /
    madhyāvalambisūtrāt talāntarāṃśās tad anyākṣaḥ //

60. Translation by Thibaut and Dvivedi, cf. *The Pañcasiddhāntikā of Varāhamihira*, ed. tr. G. Thibaut and Sudhakara Dvivedi, Varanasi 1968, 2nd edn., p. 80.
61. BORI Ms., ff. 61-62.

(bracelet[62]) and *Mudrikā* (finger- or signet-ring). All the three work on the same principle, but differ in size as their names suggest. Rāmacandra prescribes that the *Valayayantra* should measure a cubit in its diameter, the *Cūḍāyantra* a span or less, and the *Mudrikāyantra* much smaller. The inner concave surface is graduated in *ghaṭīs* for measuring time and the rim in 360 degrees for measuring altitudes.

5.6 Of the three sizes, the middle one, viz., *Cūḍāyantra*, seems to have been more popular, as the ring dials in our miniatures correspond to this size. We have also now evidence that Sawai Jai Singh, otherwise known for his gigantic astronomical instruments in masonry, used the *Cūḍāyantra* quite extensively. Before designing the masonry instruments that adorn his observatories now, he got compiled a Sanskrit manual on instruments from diverse sources, both Indian and Islamic. This text entitled *Yantraprakāra* contains not only a detailed description of the *Cūḍāyantra*, but also an elaborate set of tables to be used in conjunction with this instrument (*Cūḍāyantrasya Sāraṇī*). There are 19 separate tables, one for each decan of the zodiac, and prepared for the latitude of Delhi (28; 39°).[63]

5.7 Through a happy coincidence, Dr A.G. Kulkarni, while preparing an inventory of portable instruments and small scale models deposited in the storerooms of Jai Singh's observatory at Jaipur, located two actual specimens (one larger and the other smaller, Fig. 4.12) of the *Cūḍāyantra,* which appear to have been manufactured for Jai Singh himself.[64]

The smaller *Cūḍāyantra* has a diameter of 10.1 cm and a height of 4.9 cm. The circumference is graduated in 360 degrees. The inner concave surface is divided into 12 circumparallel columns, on

62. *Cūḍā* in the sense of a bracelet is attested from the 10th century onwards. Thus Dhanapāla's *Pāialacchīnāmamālā* (972/3 AD) 272b: *cūḍao valayabāhū; Medinīkośa* (13th c.), p. 41, v. 13ab: *cūḍā valabhau śikhāyāṃ bāhubhūṣaṇe /*

63. *Yantraprakāra of Sawai Jai Singh* (cf. n. 21), pp. 27-28, 79-82, 101, 105-114.

64. The following description of the instruments is based on the notes kindly sent by Dr A.G. Kulkarni.

which presumably the *ghaṭīs* in each solar month can be read. There are three apertures, through which the sun's rays fall upon the inner surface on the opposite side.

The larger instrument (Fig. 4.13) measures 12.9 cm in diameter and is 8.8 cm high. It has a name plate on which is engraved *cūḍāyantra dinajñāna ko*, "*Cūḍāyantra* for determining the time of the day". There are nine apertures in the breadth, and corresponding to these are nine circumparallel columns upon the inner concave surface on the side opposite the holes. The columns are graduated in *ghaṭīs.* Thus there is one aperture and a corresponding column of *ghaṭīs* for each 20 days. These columns are used successively for the first six months of the year, starting from the time when the sun enters the sign of Cancer; then in the reverse order for the second half of the year. The columns show the *ghaṭīs* in a half-day, i.e. from sunrise to midday or from midday to sunset. In the first column the numbers are from 1 to 17, which means that the duration of the day is 34 *ghaṭīs*, or 13 hrs. 36 m. This is the duration of the longest day when the sun is at the first point of Cancer. The ninth and last column has numbers 1 to 13, implying that the shortest day, when the sun is at the first point of Capricorn, comprises 26 *ghaṭīs* or 10 hrs. 24 m. These two *Cūḍāyantra* were most likely made for the latitude of Jaipur.

Thus a continuous tradition in Sanskrit astronomical texts, a series of Mughal miniatures and two actual specimens combine to provide us with a full picture of this *Cūḍāyantra*, which was hitherto entirely unknown.

5.8 In Europe, there is a similar instrument called ring dial or poke (short for "pocket"!) dial, or Bauernring in German. It is said to have been invented by George Peuerbach (1423-1461) and improved upon by his pupil Regiomontan (1436-1476). The latter also wrote a description of its construction and prepared tables of solar altitudes to be used with this instrument. The Bauernring was very popular until the last century, especially in Austria, but it is small in size like the signet ring.[65] Since the ring dial was not

65. Ernst Zimmer, *Deutsche und Niederländische Astronomische Instrumente des 11. bis 18. Jahrhunderts*, München 1979, pp. 120-122, Pl. 45.2; Charles Singer et al, *A History of Technology*, III, Oxford 1957, p. 598, Fig. 351.

known to the Islamic World, there does not seem to be any connection between the European ring dial and the Indian *Cūḍāyantra*. Thus, in this case also, we are led to conclude that the ring dial and the *Cūḍāyantra* may have developed independently in Europe and India.

6. Professional Portraits of the Astrologer/Astronomer

6.1 Having thus identified the instruments depicted in the Mughal miniatures and traced their antecedents, we can now approach some paintings where the astrologer/astronomer is portrayed along with his professional equipment. Fig. 4.8 shows an astrologer in a landscape of flowering plants, and is from the broad ornate border (*hāshiya*) of Jahāngīr's personal album. The border contains a number of professions including the astrologer. He is holding up a ring dial or *Cūḍāyantra* in his left hand taking the sun's altitude. There is a bottle of ink with a pen near his feet. An open book is in front of him, presumably a book of astronomical tables like the one compiled at the instance of Ulūgh Beg. Behind these items is a large circular disc, supported on four legs. The disc is divided into 12 segments on which the twelve zodiac figures are painted according to the Islamic fashion. What is the purpose of this zodiac board? It cannot be an instrument for time-measuring, nor for observation, nor for computation or instruction. Its only possible use is to announce that here is an astrologer ready to offer his services; in other words, it is a professional sign-board of the astrologer.[66]

6.2 There is another portrait of the astrologer (Fig. 4.10), this time quite an ironic one, in a miniature from the manuscript copy of the *Akhlāq-i Naṣīrī*. Here the astrologer is surrounded by female clientele from whom he amassed many bags of coins, and thus can afford a thick carpet, a broad sunshade and a servant to fan him.[67]

66. S.P. Verma, op. cit., p. 113, considers this to be an astrolabe. The only feature common in this object and the astrolabe is that both are circular.
67. Compare this picture with François Bernier's vivid account of the charlatanry practised by the roadside astrologers in Delhi, in his *Travels in the Mogul Empire A.D. 1656-1668,* tr. Archibald Constable, Westminster 1891, pp. 243-245.

Around him can be seen his tools of trade: a longish box for writing implements, three bags of money, a sand clock, a book stand from which he is lifting up a tome to read from it, a water pitcher, and a circular board suspended from a tripod. Goswami and Fischer identify this object as an astrolabe.[68] Like the astrolabe this disc is also surmounted by a crown (*kursī*) but unlike the former, this one is divided into 12 segments on which the zodiac signs seem to have been painted. Perhaps this too is a sign-board as in the previous painting.

6.3 While the attitude of the Mughal artist in these two paintings varies from the mildly ironic to the strongly sarcastic towards the street astrologer, he shows great reverence to another member of this profession. In Fig. 4.9, there is the portrait of a venerable old scholar in an idyllic countryside. There is a Persian wheel behind his hut and a lake in front of it with a pair of Saras cranes. He is surrounded by his disciples or mystics. There are also a number of instruments around him: a celestial globe to his proper right, a small astrolabe in front of him,[69] and also an ink bottle; to his left, close to the back wall of the hut is a small table upon which stands a sand clock, supporting an open book.

He is clearly not a professional astrologer or astronomer, but rather a saintly recluse living away from the bustle of life. What then is his connection with this assortment of astronomical instruments? Does the artist imply that these are not just measuring tools but emblems of secret knowledge which reveal cosmic mysteries to the astronomer-philosopher?[70]

68. B.N. Goswamy and Eberhard Fischer, *Wunder einer Goldenen Zeit, Malerei am Hof der Moghul-Kaiser: Indische Kunst des 16. und 17. Jahrhunderts aus Schweizer Sammlungen*, Zürich 1987, Pl. 58 and its description.
69. *A la Cour du Grand Moghol,* Paris 1986, Nr. 24, wrongly identifies this as a pendulum.
70. It is instructive to compare our painting with Hans Holbein's "The Ambassadors" (1533), where the French ambassador to England and his friend are portrayed with a number of contemporary astronomical instruments, which appear to be symbols of worldly wisdom and power.

The charm of these paintings lies in the variations they offer on the theme of the astrologer/astronomer as they show him soliciting custom on the roadside, or solemnly casting the horoscope of a prince in the royal palace, or, transcending these two mundane locations, as immersed in cosmic meditation.[71]

7. Hindu Astrologers at Akbar's Court

7.1 It will be noticed that the present exercise did not bring to light any Islamic astronomical instrument that may represent an intermediary stage of development between the observatory instruments of Maragha and Samarqand on the one hand and those of Sawai Jai Singh on the other. But then these miniature throw fresh light on three pre-Islamic Indian instruments, viz., *Ghaṭikāyantra, Kācayantra* and *Cūḍāyantra*. Though archaic in design, these instruments remained in vogue for a long span of time.

7.2 There is another likely gain. Fig. 4.3 allows us an insight into another culturally significant area. This miniature is at the Museum of Fine Arts, Boston, and depicts the birth of Salīm in 1569, who assumed the name Jahāngīr after ascending the throne in 1605. Ananda Coomaraswamy attributes this miniature to Akbar's atelier,[72] while Stuart Cary Welch opines that Jahāngīr commissioned this painting of his own birth by Bishndas.[73]

Be that as it may, the painting shows, among other scenes, four astrologers seated on a carpet just outside the red curtain covering the entrance to the harem. A lady of high birth (note her clothes and head-dress) has just brought the glad tidings of the long-awaited birth of the heir to the Mughal throne, and the astrologers have set out to measure the time of birth with the *Ghaṭikāyantra*, the sun's altitude with the *Cūḍāyantra* and to cast the horoscope.

71. This composition appealed to a later artist so much that he copied parts of it in order to create the right ambience in his "Shāh Jahān visiting a Religious Teacher". He repeats the Persian wheel, the table with the sand clock and book, and also the cranes in the foreground, but without the lake. Cf. J.V.S. Wilkington, *Mughal Painting*, London 1948, Pl. 9.
72. Op. cit. (cf. n. 3), pp. 16-17.
73. *Imperial Mughal Painting*, London 1978, pp. 70-71.

The two astrologers on either side in the foreground are Muslims and the two in the middle are Hindus, clearly distinguishable from the clothing and make-up.[74] Starting from the left, the first one in a dark green cloak is measuring the altitude with the *Cūḍāyantra*. The second, dark-complexioned, with gold lace on his turban, is the one whom the lady of the harem is addressing. He is the principal member of the panel of astrologers, which is indicated also by the fact that he is shown somewhat higher than the others. The third one, in white dress with a red shawl thrown over his shoulder, is holding in his left hand a scroll with Devanāgarī letters and with his right hand drawing the birth chart on a greyish board with chalk. The chart is drawn in north Indian Hindu fashion.[75] The fourth astrologer, wearing a dark blue shawl, is holding a paper with Persian lettering.

The dress and features of these four astrologers are so highly individualised that their pictures look like the true-to-life portraits of four prominent astrologers of Akbar's court, and it is tempting to identify them. The seniormost Muslim astrologer at Akbar's court was Maulānā or Mullā Chānd, who drew up the horoscopes of both Akbar and his son Jahāngīr.[76] Abū 'l-Fazl's glowing tribute to him

74. Apart from the Śaivite sectarian mark on the forehead, note especially how the jackets (*jāma*) are closed. The upper flap of the jacket of the Muslims runs from left to right and is fastened under the right armpit, while that of the Hindus runs from right to left. The *Akbarnāma* (III, p. 342) reports that, according to Hindu notions, "wearing the *jāma* fastened on the left side" is one of the twelve things that adorn a man. Akbar ordered that the *jāma* be tied on the right side only (*Ā'īn-i Akbarī*, I, p. 94), but this order seems to have been followed only by the Muslims. The contrary ways of closing the jacket by Muslims and Hindus correspond respectively to the modern sartorial norms of European men and women.
75. On the different modes of drawing the horoscope charts in India, see Hans Georg Türstig, *Jyotiṣa, Das System der indischen Astrologie*, Wiesbaden 1980, pp. 4-5.
76. On Maulānā Chānd, cf. *Akbarnāma*, I, p. 69; II, p. 506. He also wrote a commentary on the tables of Ulūgh Beg under the title *Tashīlāt-Mullā-Chānd-Akbar-Shāhī. Sarvadeśavṛttāntasaṃgraha*, p. 141, describes him as *jyotirvicchreṣṭhaḥ mullācāndaḥ.*

has been cited at the beginning of this article. In our miniature the elderly person on the right with white beard and blue shawl may well be the likeness of Maulānā Chānd. His colleague on the left can either be Fathullāh Shīrāzī[77] or Maulānā Alyās,[78] both of whom are reported to have prepared Akbar's horoscope.

7.3 Regarding the identity of the two Hindu astrologers in the middle, the choice is limited and likely to be more certain. In Abū-'l-Fazl's *Akbarnāma* and in Jahāngīr's memoirs, there is frequent mention of a Jotik Rāi (Sanskrit: *Jyotiṣarāja/rāya*). Abū 'l-Fazl states that Maulānā Chānd and Jotik Rāi prepared Akbar's horoscope, each according to his own tradition.[79] Jahāngīr records how Jotik Rāi's forecasts often came true.[80]

However, Jotik Rāi is not the name of an individual but an official title conferred by the Mughal emperor on the foremost

77. On this highly talented engineer and astronomer, see M.A. Alvi & A. Rahman, *Fathullah Shirazi, A Sixteenth Century Indian Scientist*, New Delhi 1968.
78. Cf. *Akbarnāma*, I, p. 126.
79. Ibid., I, pp. 85-86: "The scheme of holy nativity is hereby set down in accordance with the writing of the foremost of Indian astrologers, the Jotik Rāi, who was one of the servants of the royal household." But there was a disagreement between the two traditions: Maulānās Chānd and Alyās put Akbar's birth under Virgo and Jotik Rāi under Leo. Akbar had the horoscope of his three sons, Salīm, Murad and Dānyal, prepared by Hindu and Muslim astrologers (ibid., I, p. 85, n. 2) and again there was a discrepancy in Salīm's horoscope (ibid., II, p. 505). Interestingly, Akbar's aunt Gul-Badan Begum, in her memoirs, mentions Leo as the ascendant at the time of Akbar's birth, thus following the Jotik Rāi's horoscope (*The History of Humāyūn*, pp. 157-158).
80. *The Tūzuk-i-Jahāngīrī or Memoirs of Jahāngīr*, tr. Alexander Rogers, ed. Henry Beveridge, II, (reprinted: Delhi 1978), pp. 160, 203, 213, 235. Jahāngīr records three forecasts of Jotik Rāi that came true and refers to many of his judgements which proved to be correct. Once Jahāngīr weighed him against 6500 silver rupees and another time against 500 gold muhurs and 7000 silver rupees and gave all this money to the astrologer. As we shall see below (n. 82), Jotik Rāi gave away the money to Brahmins.

Hindu astrologer of the day.[81] Thus the Jotik Rāi of Akbar's time and the one mentioned in Jahāngīr's memoirs must be two different persons. The astrologer who made true forecasts for Jahāngīr and was weighed against silver and gold coins by the grateful monarch was Keśava, son of Kaṃhara Śarman of Kaliñjara. Keśava's son Īśvaradāśa states that his father received the title Jyotiṣarāya from Jahāngīr.[82]

Two Hindu Jyotiṣīs come into question as the likely recipients of

81. Akbar seems to have instituted the system of conferring titles with the suffix *–rāya/rāi* to various types of Hindu scholars, and his successors continued the system. Given below are the few cases that I occasionally came across:

Emperor	Title	Recipient
Akbar	Kabi Rāi	Mahesh Das *alias* Birbal
	Jotik Rāi	(Nīlakaṇṭha ?)
Jahāngīr	Jotik Rāi	Keśava, s.o. Kaṃhara Śarman
Shāh Jahān	Vedāṅgarāya	Mālajit, author of *Pārasīkaprakāśa*
	Paṇḍitarāya	Jagannātha, author of *Rasagaṅgādhara*, etc.
	Kabi Rāi	Sundar Das, a Hindi poet from Gwalior
Muhammad Shāh	Jyotiṣarāya	Kevalarāma, associate of Sawai Jai Singh
	Paṇḍitarāya	Nayanasukha, associate of Sawai Jai Singh
Bahādur Shāh	Jotik Rāi (?)	Sukhanand, court astrologer

82. In the colophon to his *Muhūrtaratna*, which he completed in 1663, he gives the following information about his father Keśava:

tasmāt keśavaśarmābhūt khyātas triskandhavikramaḥ /
gajāśvarathasaṃgrāmair yo 'rcitas syān nṛpottamaiḥ //
so 'yaṃ jyotiṣarāyākhyaḥ jyahangīrāvanīpateḥ /
svagūḍhapraśnasaṃvādair lebhe praśnavidāṃ varaḥ //
tulāpramukhadāneṣu yas tv asaṃkhyavasūni vai /
viprasāt kṛtavān kāle nārāyaṇaparāyaṇaḥ //

cited in M.M. Patkar, "Muhūrtaratna: A religio-astrological treatise, composed in the Reign of Aurangzeb," *The Poona Orientalist,* 3 (1938-39), pp. 8-85, esp. 85. See also, *CESS*, A-3, pp. 55-56.

the title from Akbar: Nīlakaṇṭha[83] and Kṛṣṇa Daivajña.[84] They were not only the eminent astrologers of the time; they were also mediators between the Hindu and Muslim traditions of astronomy and astrology.

7.4 Nīlakaṇṭha was a protégé of Akbar's minister Ṭoḍarmal. In 1572 he wrote the *Jyautiṣa-saukhya* and some other sections of the *Ṭoḍarānanda*, an encyclopaedic compendium prepared at the instance of Ṭoḍarmal.[85] In 1587, Nīlakaṇṭha composed the *Tājikanīlakaṇṭhī*, a well known manual on Islamic astrology.[86] His son Govinda wrote a commentary on it in 1622, which he styled simply *Rasālā* (from Arabic *risālah*, "tract"). Elsewhere Govinda states that his father was an incomparable ornament of Akbar's court.[87] Govinda's son Mādhava, who also wrote a commentary on the grandfather's *Tājikanīlakaṇṭhī*, proudly proclaims that his

83. Ibid., A-3, pp. 177-189.
84. Ibid., A-2, pp. 53-55.
85. Cf. *Ṭoḍarānanda*, I, ed. P.L. Vaidya, Bikaner 1948, pp. 396-404.
86. David Pingree, *Jyotiḥśāstra: Astral Mathematical Literature*, Wiesbaden 1981, pp. 98-99. Abū 'l-Fazl reports that "At the command of his Majesty, Mukammāl Khān of Gujarāt translated into Persian the Tājak, a well known work on Astronomy" (*Ā'īn-i Akbarī*, I, p. 112). *Tājaka/tājika* is not the name of a book on astronomy but refers to a class of Sanskrit works on astrology which are Indian adaptations of Arabic/Persian astrology. The most famous of such works is the *Tājikanīlakaṇṭhī*. Is this the one that Akbar got translated into Persian? But it would indeed be strange to get a contemporary work on Arabic /Persian astrology translated again into Persian. Sir William Jones emends the word *tājak* into *jātak* (cf. Beveridge, *Akbarnāma*, I, p. 91n) and it is more likely that a Persian translation was commissioned of Varāhamihira's *Bṛhajjātaka* or some other popular manual on Indian horoscopy. However, manuscript copies of neither text in Persian rendering are listed in M. Habibullah, "Medieval Indo-Persian Literature relating to Hindu Science and Philosophy, 1000-1800 A.D.," *Indian Historical Quarterly*, 14 (1938), pp. 167-181.
87. In his commentary called *Pīyūṣadhārā* (1603) on the *Muhūrtacintāmaṇi* (1600) of his paternal uncle Rāma, Govinda says the following about his father Nīlakaṇṭha:

grandfather Nīlakaṇṭha was honoured by Akbar and that his father Govinda was honoured by Jahāngīr.[88] I would like to imagine that the dark-complexioned astrologer with the Śaivite mark on his forehead and gold lace in his turban in our miniature is a true likeness of this Nīlakaṇṭha and that he held the title and office of Jotik Rāi at Akbar's court.

7.5 The other Jyotiṣī who is drawing the birth chart in the miniature then must be Kṛṣṇa Daivajña, son of Ballāla. His younger brother Raṅganātha states that Kṛṣṇa was honoured by Jahāngīr.[89] Raṅganātha's son Munīśvara also says that Kṛṣṇa was a favourite of Jahāngīr.[90] Kṛṣṇa wrote an excellent commentary on Bhāskara's

---

sīmā mīmāṃsakānāṃ kṛtasukṛtacayaḥ karkaśas tarkaśāstre
jyotiḥśāstre ca gargaḥ phaṇipatibhaṇitivyākṛtau śeṣanāgaḥ /
pṛthivīśākabbarasya sphuradatulasabhāmaṇḍanaṃ paṇḍitendraḥ
sākṣāc chrīnīlakaṇṭhaḥ samajani jagatīmaṇḍale nīlakaṇṭhaḥ //

Cf. *Muhūrtacintāmaṇi* with the commentary *Pīyūṣadhārā* of Govinda, Bombay 1946, pp. 1-2 (verse 8 at the beginning of the commentary).

88. Thus he says of Nīlakaṇṭha at the beginning of his commentary *Śiśubodhinī:*

triskandhaṃ jyotiṣaṃ ca samakṛta viśadaṃ ṭoḍarānandasaṃjñam /
sāhityajñānapūrṇaḥ smṛtiṣu ca nipuṇo 'kabbarakṣmeśamānyaḥ //

and of his father Govinda that he was *nṛpativarajahāṅgīraśāhātimānyaḥ /*

Cf. *CESS,* A-4, pp. 415-417. Govinda's association with Jahāngīr's court was mentioned also by his other son Cintāmaṇi in his commentary on Raghunātha's *Muhūrtamālā,* written in 1661 during the reign of Aurangzeb (*rājye 'varaṅgajevasya*), cf. ibid., A-3, pp. 49-50.

89. At the conclusion of his commentary called *Gūḍhārthaprakāśa* on the *Sūryasiddhānta* (many editions), he describes his elder brother Kṛṣṇa thus:

tataḥ sa kṛṣṇo jahāṃgīrasārvabhaumasya sarvādhigatapratiṣṭhitaḥ /
śrībhāskarīyaṃ vivṛtaṃ tu yena bījaṃ tathā śrīpatipaddhatiś ca //

This was written in 1603, i.e. two years before Prince Salīm became the *sārvabhauma* and assumed the name Jahāngīr (cf. *Tūzuk-i Jahāngīrī,* I, pp. 2-3). Another case of forecast that came true?

90. About his uncle Kṛṣṇa, Munīśvara states the following in his commentary on the *Siddhāntaśiromaṇi* of Bhāskarācārya:

Bījagaṇita.[91] In Akbar's translation bureau, where several representative Sanskrit works were translated into Persian and vice versa, Ulūgh Beg's astronomical tables were translated into Sanskrit by the joint efforts of Muslim and Hindu scholars. The Muslims in the group were Fathullāh Shīrāzī and Abū 'l-Fazl and the Hindus included Kṛṣṇa.[92] Kṛṣṇa also wrote a commentary on Śrīpati's *Jātakapaddhati*, a manual on preparing horoscopes.[93] In this commentary, Kṛṣṇa included the horoscope of Khān-i-Khānān 'Abd al-Rahīm Khān, who was an influential courtier of Akbar, at one time tutor of Salīm, and himself a famous man of letters.[94] It is therefore appropriate that Kṛṣṇa should be portrayed in our miniature as drawing the horoscope.

---

yaḥ śrīkṛṣṇapadāmbujahitamatiḥ siddhāntavārāṃnidheḥ
potaḥ śrījahagīrabhūmitilakasyānanaviśvāsabhūḥ /
ṣaṭśāstreṣu kṛtaśramo 'khilagurur mānyo vadānyo vidāṃ
śrīkṛṣṇaḥ kim u varṇanīyavibhavaḥ śrīkṛṣṇa evāparaḥ //

Cf. M.M. Patkar, "Moghul Patronage to Sanskrit Learning," *Poona Orientalist*, 3 (1938-39), pp. 164-175, esp. 169.

91. *Bījagaṇita* of Bhāskara with the commentary *Navāṅkurā* of Kṛṣṇa Daivajña, ed. V.G. Apte, Poona 1930.
92. Cf. M.A. Alvi & A. Rahman, *Fathullah Shirazi*, p. 24: "A part of the *Zīji-i Jadīd-i Mīrzāi* (astronomical tables of Ulugh Beg) had been translated under his guidance [i.e. Fathullah Shirazi] by Kishan Jotishi, Ganga Dhar, Mahesh Mahanand and Abul Fazl." This information stems from the *Ā'īn-i Akbarī*, but Blochmann's translation (I, p. 110) is hopelessly garbled. The Maharaja Sawai Man Singh II Museum, Jaipur, possesses the unique copy of Ulūgh Beg's tables in Sanskrit rendering, entitled *Jīca Ulūgabegī* (Ms. no. 45). I have briefly examined it to see if the rendering stems from Akbar's court. Except on the last page, there is no text but only tables. On the last page, the method of finding out the weekday corresponding to a given date and the like are explained in Rājasthānī, thus suggesting that this version may have been prepared anew for Sawai Jai Singh.
93. *Jātakapaddhatyudāharaṇa*, ed. Jatindra Bimal Chaudhuri, Calcutta 1955.
94. 'Abd al-Rahīm Khān himself wrote a small tract on *Tājika* astrology in Sanskrit under the title *Kheṭakautuka*, in which he sprinkles Arabic/Persian technical terms. There are several editions of this work, cf. *CESS*, A-2, pp. 79-80.

7.6 Even if my identification hits the mark only in parts, there can be no doubt that here we have a unique group portrait of the prominent scientists of Mughal India. Moreover, this assemblage of Hindu and Muslim men of learning in a painting commissioned by the emperor reflects Akbar's policy of encouraging the scientific activity of both traditions and of promoting their synthesis. This miniature then is a graphic representation of the interaction between the Hindu and Islamic traditions of scientific knowledge.

Acknowledgements

This is an expanded version of an invited lecture delivered at the Symposium on Asian Science, Medicine and Technology, during the XVIIIth International Congress of History of Science, held at Hamburg and Munich in August 1989. I should take this opportunity to thank Professors C.J. Scriba (Chairman of the Organising Committee of the Congress) and S.M.R. Ansari (Organiser of the Symposium) for inviting me. I am grateful to Professors E.S. Kennedy (Princeton) and David A. King (Frankfurt) for their advice on Islamic astronomical instruments; to my colleagues Professor A.J. Qaisar and Dr S.P. Verma for patiently guiding me through the treasures of Mughal miniatures. Professor Qaisar also generously let me use his personal collection of slides and photographs. Cordial thanks are due to Drs Nalini Balbir (Paris) and Bettina Bäumer (Varanasi) for sending photographic reproductions; to Shri Hazari Mull Banthia (Kanpur) and Mr Yukio Ohashi (Tokyo) for xerocopies of material inaccessible to me; Dr A.G. Kulkarni (Bangalore) for the photos and information on the *Cūḍāyantras* at Jaipur; and to my young friend Sudhir Singh for photography.

Abbreviations Used

Ā'īn-i Akbarī: The Ā'īn-i Akbarī by Ābū'l Fazl 'Allāmī. vol. I, tr. H. Blochmann; revised by D.C. Phillot. Reprint: Delhi 1977; vols. II, III. tr. H.S. Jarrett, revised and annotated by Jadunath Sarkar. Reprint: Delhi 1978.

Akbarnāma: The Akbarnāma of Abu-l-Fazl (*History of the Reign of Akbar including an Account of his Predecessors*), tr. H. Beveridge, 3 vols., Calcutta 1910.

CESS: David Pingree, *Census of the Exact Sciences in Sanskrit,* Series A, vols. 1-4, Philadelphia 1970-1981.

Qaisar: Ahsan Jan Qaisar, *The Indian Response to European Technology and Culture (A.D. 1498-1707),* Delhi 1982.

Turner: A.J. Turner, *The Time Museum, Catalogue of the Collection,* vol. I, part 1: *Astrolabes, Astrolabe-related Instruments*; part 3: *Water-clocks, Sand-glasses, Fire-clocks*, Rockford 1984-85.

Fig. 4.1. "Birth of Murad" by Bhurah and Baswan, ca. 1600, from the *Akbarnāma.* Victoria & Albert Museum, London, IS 2-1896, no. 80/117. Cf. S.P. Verma, *Art and Material Culture in the Paintings of Akbar's Court*, New Delhi 1978, Pl. XIII.

Fig. 4.1A. "Astrologers casting the horoscope", detail from Fig. 4.1.

Fig. 4.2. "Astrologers explaining the horoscope to the king", detail from the "Birth of Akbar", *Akbarnāma*, British Library, London, Ms. Or. 12988, f. 20b. Cf. Geeti Sen, *Paintings from the Akbar Nama: A Visual Chronicle of Mughal India,* Calcutta 1984, Pl. 57, pp. 130-131.

Fig. 4.3. “Astrologers casting the horoscope”, detail from the “Birth of Salīm”, Museum of Fine Arts, Boston, 17.3112. Cf. Stuart Cary Welch, *Imperial Mughal Painting*, London 1978, Pl. 16, pp. 70-71.

Fig. 4.4. "Astrologers casting the horoscope", detail from the "Birth of Salīm", Chester Beatty Collection, Dublin. Cf. Hermann Goetz, *India: Five Thousand Years of Indian Art*, Bombay 1959, pp. 212-213.

Fig. 4.5. "Astrologers casting the horoscope", detail from the "Birth of Timūr", *Akbarnāma*, British Library, London, Ms. Or. 12988, f. 34b. Cf. A.J. Qaisar, *The Indian Response to European Technology and Culture* (*A. D. 1498-1707*), Delhi 1982, Pl. 10.

Fig. 4.6. "Jahāngīr seated on an hourglass", by Bichitr, ca. 1625. Freer Gallery of Art, Smithsonian Institution, Washington, 45.15a. Cf. Stuart Cary Welch, *Imperial Mughal Painting,* London 1978, Pl. 22, pp. 82-85.

Fig. 4.7. "Angels holding a crown, celestial globe and ring dial", detail from a portrait of Humāyūn, Shāh Jahān's Album, ca. 1650. Sackler Gallery of Art, Smithsonian Institution, Washington. Cf. Constanze A. Bond, "A Priceless Collection Rediscovered", *SPAN*, XXX.2 (May 1989).

Fig. 4.8. "The Astrologer" from the border of Jahāngīr's Album, Náprstek Museum, Prague. Cf. Lubo Hajek, *The Indian Miniatures of the Mughal School*, London 1960, Pl. 18.

Fig. 4.9. "Astronomer with his disciples", Shāh Jahān's Album, Musée Guimet, Paris, MA 2471. Chliché des Musées Nationaux-Paris.

Fig. 4.10. "Bazar astrologer with his clients", detail from a manuscript illustration to the *Akhlāq-i Naṣīrī*, ca. 1590-1595. Prince Sadruddin Aga Khan Collection, Museum Rietberg, Zürich, cf. B.N. Goswamy & Eberhard Fisher, *Wunder einer Goldenen Zeit. Malerei am Hof der Moghul-Kaiser: Indische Kunst des 16. und 17. Jahrhunderts aus Schweizer Sammlungen*, Zürich 1987, Pl. 58.

Fig. 4.11. "Noah's Ark", Freer Gallery of Art, Smithsonian Institution, Washington. Cf. Stuart Cary Welch, *Imperial Mughal Painting*, London 1978, Pl. 9.

Fig. 4.12. Two *Cūḍāyantras* from Jai Singh's Observatory at Jaipur. Courtesy Directorate of Archaeology and Museums, Government of Rajasthan.

Fig. 4.13. The larger *Cūḍāyantra* from Jai Singh's Observatory at Jaipur. Courtesy Directorate of Archaeology and Museums, Government of Rajasthan.

PART II

The Water Clock

5

The Bowl that Sinks and Tells Time

About the middle of the twelfth century, Bhāskarācārya, the celebrated mathematician and astronomer, wrote a book on arithmetic, to which he gave the tantalising name *Līlāvatī.* The name literally means the 'graceful one' and is given usually to girls. Why then did Bhāskara give such a name to his book which, though composed in charming verse, dealt with a very dry subject?

The book was so popular that 400 years later, Akbar chose it as one of the several representative Sanskrit texts to be translated into Persian for the Muslim intelligentsia to gain a better appreciation of Hindu learning. The poet Shaikh Abū 'l-Faiz Faizī, who was entrusted the task of rendering the *Līlāvatī* into Persian, was intrigued by the title. Probably his Brahmin assistants narrated the following story which he reproduced in his preface to the translation which was completed in 1587.

Līlāvatī, so runs the story, was the name of Bhāskara's daughter. When she was born, the astrologers predicted that she would never wed. Bhāskara, however, divined a lucky moment for the marriage and left an hour cup floating on the vessel of water. The cup had a small hole at the bottom so that the water would trickle in and it would sink at the end of that hour. However, when Līlāvatī, curiously came to see the water rising in the cup, a pearl from her garment dropped into the cup, stopping the influx. So the hour passed without the sinking of the cup, and Līlāvatī never married. To console her, Bhāskara wrote a book in her honour, saying "I shall write a book of your name which shall remain to the latest time; for a good name is a second life and the ground-work of eternal existence."

First published in: *India Magazine*, 14.9 (September 1994) 31-36.

Whether true or not, this story refers to a time-measuring device that was in vogue both at the time of Bhāskara and Akbar. Indeed, it has been in use throughout South Asia and beyond for over fifteen hundred years up to the turn of the present century. Even now, there are isolated cases of its employment for determining the time of certain rituals.

This device, technically known as the sinking bowl type, is quite different from the water clocks, or clepsydrae, used in ancient Egypt, China or the Roman empire. This one consists of a hemispherical bowl, made of copper, with a minute aperture at the bottom. When this bowl is placed on the surface of water in a larger basin, water enters the bowl through the aperture below. As soon as the bowl becomes full, it sinks to the bottom of the basin with a clearly audible thud. The weight of the bowl and the size of the aperture are so adjusted that the bowl sinks sixty times in a day-and-night. Thus the duration of each immersion is 24 minutes, which has been the standard unit of time in India. In Sanskrit, the bowl is called *ghaṭī* or *ghaṭikā* (literally, small pot) and these two terms also designate the length of time measured by this device. The whole apparatus was called *Ghaṭī-yantra* or *Ghaṭikā-yantra.*

Starting from Āryabhaṭa at the end of the fifth century, many astronomers have described this device. Perhaps the most detailed description appears in Viśrāma's *Yantraśiromaṇi* written in 1615:

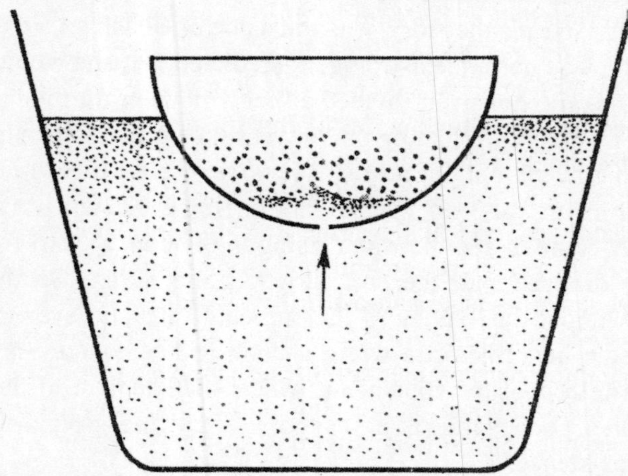

Fig. 5.1. Working principle of the sinking bowl type of water clock.

> Now for the sake of the welfare of the virtuous, I teach the method of manufacturing the water clock for determining the exact time. With ten palas of copper, make a hemispherical bowl. It should be well-rounded, have a diameter of twelve *aṅgulas* and hold sixty palas of water. Next draw a wire from three and half *māṣas* of gold so that it is uniformly round and four *aṅgulas* long. With this wire, the wise man, after due reflection, should bore a hole at the exact centre of the bowl. Then the bowl should be gilded and placed in a basin full of water. The time of immersion is considered to be one *ghaṭī*. From one sunrise to the next, if the bowl sinks again and again sixty times, then it is accurate. If the number of immersions is less or more, the wise man should correct the water clock by enlarging the hole or by reducing it through appropriate technical means. When the measurement is made in a windless place and with pure water, the *ghaṭī* will be accurate.

It is likely that the original inspiration for the sinking bowl came from the coconut shell, naturally endowed with a hole. In south India, water clocks made of well-scrubbed halves of the coconut shell survive in museums. The first mention of this kind of water clock occurs in a commentary written by Buddhaghoṣa in the first half of the fifth century in Sri Lanka, which is a coconut-producing country. Therefore, it is reasonable to suppose that the sinking bowl type of water clock was invented some time in the fourth century either on the southern coast of India or in Sri Lanka, or on one of the islands of the Malay Archipelago.

By the first half of the fifth century, it had become the standard equipment for measuring time in Buddhist monasteries in Sri Lanka. In his commentary on the *Majjhimanikāya*, Buddhaghoṣa, renowned for his commentaries of Buddhist scriptures, narrates the story of a monk, whose duty it was to measure time by means of the sinking bowl and announce it by striking the bell. The monk, appropriately called Kāladeva, the Lord of Time, became so adept that he did not need to watch the actual sinking of the bowl. When the bowl sank, his hand automatically lifted the mallet to strike the bell.

In India, the earliest mention of the sinking bowl occurs in the writings of Āryabhaṭa at the end of the fifth century. From the seventh century onwards, institutions for time-keeping are attested at royal palaces, Buddhist monasteries, temple courtyards, town squares and so on, where time was measured with the sinking bowl and announced by diverse instruments, such as bells, drums, conch shells and gongs.

Two kinds of time units were measured and announced in these establishments. The first is, of course, the *ghaṭī*, the time taken for the bowl to sink. The *ghaṭī* is always of uniform duration, twenty four minutes, and is divided into sixty *vighaṭīs* or *palas*, which equal twenty four seconds. One can, if one so wishes, divide and mark out the inner surface of the bowl empirically, if not into sixty, at least into ten equal parts and thus measure smaller units of six *palas* or two minutes and twenty four seconds. The *ghaṭīs* and *palas* were employed primarily for astronomical and astrological purposes.

In other endeavours, larger units of time would suffice. Here the duration of daylight, from sunrise to sunset, is divided into four quarters called *praharas*. The night also had four quarters called *yāmas*. However, since the duration of daylight varies according to seasons, the lengths of these *praharas* and *yāmas* also varied. At equinoxes, when the day and night are equal and measure exactly thirty *ghaṭīs* (or twelve hours), the *prahara* and *yāma* are equal and measure seven and half *ghaṭīs* (or three hours). There have been various systems for reconciling the variable *praharas* with the fixed *ghaṭīs*, but it will be too tedious to go into them here. Suffice it to say, that in all time-keeping establishments, not only the passage of each *ghaṭī* was announced, but also the completion of each *prahara* and *yāma*.

Sometimes, two kinds of instruments were used to denote *praharas* and *ghaṭīs* respectively. Thus in the *Kādambarī*, the Sanskrit romance composed by Bāṇa in the seventh century, the passage of a *ghaṭī* was announced in the royal palace by drum beats, but the completion of the second *prahara* since sunrise, that is midday, by a loud blast on the conch shell. Upon hearing this, the king broke up the audience and proceeded towards the private apartments for his bath and meal.

But more often, both *praharas* and *ghaṭīs* are indicated by the same instrument, the bell in south India and the drum in the north. In order, however, to distinguish between the two kinds of strokes, the time-keeper executes a series of rapid strokes between the *prahara*-strokes and the *ghaṭī*-strokes. To take a concrete example, if it is now two *praharas* and five *ghaṭīs* since sunrise, this will be announced in the following manner: two clear strokes, pause, a series of rapid strokes, pause, five distinct strokes.

These time-keeping establishments naturally required a large number of personnel. At least two men must attend the water clock constantly, one to empty the bowl as soon as it sinks and to set it afloat again, the other to strike the *ghaṭīs*. These two men are relieved by another pair after each *prahara* or *yāma*. Thus these words are analogous to the English word 'watch'. Therefore, only kings could afford such establishments in order to regulate their numerous engagements. Monasteries and temples with rich endowments also maintained these establishments for performing the various religious observances according to a fixed schedule.

In the last quarter of the seventh century, a Chinese Buddhist pilgrim by name I-Tsing came to India. He spent ten years at the monastery of Nalanda and also visited those at Gaya and Kuśinagara. In all these places, he found time-keeping etablishments, endowed by generations of kings. He concludes his detailed account of these establishments by relating the advantages of the water clock over the sundial and recommending its use in the monasteries of China as well:

> Owing to the use of those clepsydrae (water clocks), even in thick clouds and in a dark day, there is no mistake whatever about the horse-hour (noon) and even when rainy nights continue, there is no fear of missing the watches. It is desirable to set such ones (in the monasteries in China) asking for royal help, as it is a very necessary matter among the Brotherhood.

At the beginning of the eleventh century, al-Bīrūnī wrote about similar time-keeping establishments at Peshawar, for the regular maintenance of which pious people had bequeathed fixed incomes. This system prevailed in Gujarat also. An early medieval document *Lekhapaddhati* lists, among the various administrative departments, the *ghaṭikā-gṛha-karaṇa* which obviously was a department to oversee the time-keeping establishments in the Chaulukya-Vāghela kingdom. The establishments themselves are mentioned in inscriptions and other documents under the names *ghaṭikā-gṛha, ghaṭī-gṛha,* or *ghaṭikālaya* (all meaning water-clock-houses). Finally, in a painted book cover from about 1130, there is a unique picture of a water-clock-house, situated next to the Nemināthа temple at Āśāpallī.

During the twelfth or the thirteenth century, the drum and conch shell were replaced by the metal gong to announce time in northern India. This gong was called *ghaḍiyāl* or *ghariyāl*, words phonetically derived from *ghaṭikālaya* (water-clock-house). The new ensemble of the bowl and gong are mentioned for the first time, under the Persian name *ṭās-i ghariyāla*, by Shams-i-Sirāj 'Afīf in his biography of Sulṭān Fīrūz Shāh Tughluq. Fīrūz, who ruled Delhi from 1351 to 1388, was very interested in sciences and engineering and ushered in many technical innovations. He installed the *ṭās-i ghariyāla* at the top of his palace gate at Firuzabad, as Hindu kings had done previously for several centuries.

The Mughal emperor Bābur was also highly impressed by the system of time-keeping with the bowl and gong. He not only introduced the system at his court but also made improvements in the mode of announcing the *pahars* and *gharis* at night. More important still is his account in his memoirs, where he speaks of a professional body of people called *ghariyālīs* who maintained the water clocks and announced the time in every major town of Hindustan. This time-keeping system continued under Akbar also.

Fig. 5.2. *Ghaṭī-yantras* at the Pitt Rivers Museum, Oxford.
This is the largest and the most representative collection containing specimens in diverse shapes from Malabar, Tamil Nadu, Uttar Pradesh and also Sri Lanka and Burma. While six are of copper, the one in the foreground was fashioned from a coconut.

Fig. 5.3. Emperor's time piece from Burma. Copper water clock, called *nayi*, from the Emperor's palace at Mandalay, Burma. Pitt Rivers Museum, Oxford. (Acc. No. 1892.41.37.1). The bowl is rather large with an unusual near-cylindrical shape. Diameter at the base 132 mm, at the mouth 149 mm, height 105 mm. The duration of immersion is not recorded. To the right of the bowl are eight large bone beads strung in a wire for counting the hours as they are completed.

The profession of the royal time-keeper was apparently a hazardous one. J. Nisbet reports thus in his *Burma Under British Rule, and Before* (1901):

As each *nayi* was measured off, a gong was beaten, and at every third hour the great drum-shaped gong was sounded from the *pahozin* or time-keeper's tower within the inner precincts of the royal palace at the eastern gate. From the *paho* the beats were repeated on large bells by all the guards throughout the palace. To ensure attention to this matter in the olden days, the time-keeper could be carried off and sold in the public market, if he were negligent in the discharge of his duties, being then forced to pay a fine in the shape of ransom.

Fortunately, in two miniature paintings produced at his atelier, astrologers are shown measuring the time of a royal birth by means of the sinking bowl and drawing the horoscope. These are the only pictorial representations of this device (see Figs. 4.2 and 4.3)

As the Mughal empire declined and minor nobles began to put on imperial airs, they too installed the bowl and the gong at the gates of their palaces and fortresses. Between the seventeenth and nineteenth centuries, several European visitors wrote about the sinking bowl and the mode of announcing time with its help. These accounts, often inaccurate and full of outlandish transcription of Indian names, do not add much to our knowledge, except to show that this archaic time measurement continued without much challenge from the European clocks. Occasionally, of course, some bowls were made to measure the European hours of sixty minutes or half hours. The best of such European accounts is from Meer Hassan Ali's English wife, who lived in Lucknow with her husband's family for twelve years from 1816. In her *Observations on Mussalmauns in India*, Mrs. Meer Hassan Ali shows how the sinking bowl had become an integral part of the upper class Muslim households at the beginning of the nineteenth century. After describing how the sinking bowl is used to measure time, she goes on to say:

> Each hour, as it passes, is struck by the man on duty with a hammer on a broad plate of bell-metal, suspended to the branch of a tree, or to a rail: the gong of an English showman at the country fairs is the exact resemblance of the metal plates used in India for striking the hours on, and must, I think, have been introduced into England from the East.
>
> The *durwan* (gate-keeper), or the *chokeedars* (watchmen), keep the time. In most establishments, the watchmen are on guard two at a time, and are relieved at every watch, day and night. On these men devolves the care of observing the advance of time by the floating vessel, and striking the hour, in which duty they are required to be punctual, as many of the Mussalmauns' services of prayer are scrupulously performed at the appointed hours.

Hindu rajas, maharajas and jamindars had *jalgharis* (water clocks) or *katoris* (bowls) installed at the gates of their kothis and had the time announced regularly. But common householders did not need and could not afford them. However, they too had the water clock set up in their homes on special occasions like marriages in order to know the exact moment which is astrologically favourable for the marriage. The priest who performed the marriage usually brought the bowl with him and set it up at the client's house with due ceremony and chanted a prayer such as this one:

> The Creator made you long time ago
> As the best of the time-telling devices,
> For increasing the life of the married pair,
> And for bestowing sons, wealth and the like on them.
> Grant me, therefore, O little water clock,
> This day the fulfillment of all my desires.

The story of Līlāvatī refers to this very practice of setting up a water clock in private homes just for the day of the marriage. In south India, apparently the bridegroom was to look after the clock and literally count the *ghaṭīs* up to the auspicious moment. Pillalamarri Pinavīrabhadra, a fifteenth century Telugu poet, alludes to this tradition in a charming invocation at the beginning of his *Jaiminibhāratamu.* When Śiva came to wed Pārvatī, says the poet, he carried on his head, as he always did, the half moon and Gaṅgā. To Pārvatī's maids, the half moon looked like the bowl afloat on the water of Gaṅgā. So they teased Pārvatī saying: 'Look at your husband-to-be. He is so impatient for the marriage that he is carrying the water clock on his own head'.

By the turn of this century, the sinking bowl gradually yielded its place to the pendulum clock, but even then, there were occasions when it was considered more reliable. When Rudyard Kipling's father, J.L. Kipling, was the curator of Lahore Museum, the modern clock was out of order so often that he installed a water clock for the use of the policemen guarding the museum. In the town of Sehwan in Sindh, Pakistan, there is the mausoleum of Shaikh 'Usmān, popularly known as Lāl Shahbāz Qalandar. At this mausoleum, the sinking bowl was in use as late as 1973 for regulating the times of various daily rites. The practice may still be continuing. According to Professor N.A. Baloch, the bowl must have been in use during the lifetime of the saint in the thirteenth century for determining the times of prayer and of the dance of the dervishes (*dhammal*) just after the Maghrib prayer. After the saint's death, the ceremonies which became more elaborate were still performed according to the time measured by the water clock. In Sindhi, the bowl is called *wato*, the water basin in which it is made to float is known as *degrro*, and the gong *gharryal.*

And in this year 1994, at Jhalawar, a 'dusty little place, 87 kilometres south of Kota' in Rajasthan, well known travel writers Hugh and Colleen Gantzer, saw a floating water clock in use at he Jaina shrine dedicated to the Tīrthaṅkara Śāntinatha, and wondered: 'We thought that floating water clocks were a Chinese invention. Had Jhalawar any contact with China? Or was this an Indian invention?' (*The Hindu*, 24 April 1994).

Thus the technology of bygone ages becomes at times fossilised in the ritual, like the fly in amber, and is preserved for posterity. Language too, quite often, retains a memory of things. In north Indian languages, all time-keeping devices, however sophisticated they might be, are still called *ghaḍī*, which is just a phonetic variation of *ghaṭī*, the little pot that sank and told time. The corresponding south Indian names *ghaḍiyāramu* and *gaḍigāram*, are derived from the Sanskrit *ghaṭikāgāra*, the water-clock-house. Finally, when we say 'it struck four' or something similar, the expression is not derived from the chimes of the European clocks but from the old practice of striking the *ghaṭīs* and *praharas* on the gong or the bell.

During the last fifteen hundred years, since the sinking bowl was first mentioned in India, many varieties of time-measuring instruments were invented in India and also imported from outside. Some of these, like the *dhruvabhrama-yantra* invented by Padmanābha around 1400 or the astrolabe imported from the Islamic World, were highly sophisticated and could be used both in the daytime and at night, but none could replace the archaic bowl which held sway throughout India and surrounding countries. Hindus and Muslims, Jainas and Buddhists, used it for both religious and secular purposes. In view of its universal popularity, there ought to be hundreds of specimens in every part of India and also in museums abroad.

In order to preserve these cultural and scientific artifacts for posterity, I recently began to compile a catalogue of all existing specimens of Indian astronomical and time-measuring instruments, whether deposited in India or outside. In this connection, I have seen all the known specimens in the USA and UK. The largest collection of sinking bowl water clocks is at the Pitt Rivers Museum at Oxford. There are seven bowls from different regions: Malabar, Tamil Nadu, Mirzapur, Burma and Sri Lanka. The sizes and shapes are all different as are the materials. Six are made of copper while the seventh one from Malabar is made from a coconut shell. There is one more copper bowl at the Horniman Museum at Forest Hill, a suburb of London, and another at the Worshipful Company of Clockmakers' Collection in the Guildhall Library at London. These nine pieces constitute a reasonably representative collection in UK.

In India, I have not yet begun a systematic survey of museums and private collections. Therefore, the water clocks I have seen or heard about are not many. In 1907, Edgar Thurston, Superintendent of the Madras Government Museum, wrote that there were several specimens in his museum. Today there is only one, made of coconut shell. My friend Professor S. Ramaratnam of the Vivekananda College, let it float in a bucket of water and found that it measured exactly twenty-four minutes. I have seen copper bowls in the Archaeological Museum at the Golconda Fort, Hyderabad, and in the Museum of Indology, Jaipur. The palace museums at Bharatpur and Udaipur are said to have one each. There ought to be many more than this and I shall be grateful for information about any new specimens.

It is a matter of great regret that not many specimens of this unique cultural document are preserved. Even amongst the few that are preserved, there is not a single one that has been graduated to measure submultiples of *ghaṭī*. Regrettably, no basins seem to be extant in which the bowl was set afloat, not to speak of the gongs and bells on which time was broadcast. Unless anthropologists and local museums systematically begin to collect artifacts of daily use like these, an important aspect of India's past will be irretrievably lost.

ACKNOWLEDGEMENTS

For allowing me to study their collections, grateful thanks are due to Ms. Linda Mowat, Assistant Curator, Pitt Rivers Museum, Oxford; and to Sir George White, Curator, the Worshipful Company of Clockmakers' Collection, The Guildhall Library, London.

6

Announcing Time: The Unique Method at Hayatnagar, 1676

1.1 The chief device for measuring time in India from about the fourth century AD up to the beginning of the present century has been a water clock of the type known as the "sinking bowl".[1] This type of water clock consists of a hemispherical bowl, made of thin copper, with a minute perforation at the bottom. When this bowl is placed on the surface of the water in a larger vessel, water enters the bowl from below through the perforation. As soon as the bowl is full, it sinks to the bottom of the vessel. The perforation is so made that the duration of each sinking of the bowl is 24 minutes, or one-sixtieth part of the nychthemeron, this being the standard unit of time in India.

1.2 This apparatus is first mentioned by Buddhaghoṣa, in the early fifth century AD, in his commentary *Papañcasūdanī* on the *Majjhima-nikāya.*[2] Subsequently, astronomers like Āryabhaṭa (6th c.), Varāhamihira (6th c.), Lalla (8th c.) and Bhāskara (12th c.) described it among other time measuring and astronomical instruments. Some of these writers gave also precise measurements for the water clock. Thus, for example, Āryabhaṭa laid down that

Jointly with Ishrat Alam. First published in: *Proceedings of the Indian History Congress, 52nd Session, New Delhi 1991-92,* Delhi 1992, pp. 426-31.

1. On the types of water clocks, see Joseph Needham, *Science and Civilisation in China,* III, Cambridge 1959, p. 315; A.J. Turner, *The Time Museum: Catalogue of Collections*, I, Part 3, Rockford 1984, p. 1.
2. *The Papanchasudani*, the Commentary on the *Majjhima-Nikaya*, vol. I, ed. U. Dhammaratana & U. Jagarabhivamsa, Nalanda 1975, p. 158. Cf. also, Oskar von Hinueber, "Probleme der Technikgeschichte im alten Indien", *Saeculum* 29.3 (1978), p. 224 and n. 44.

the bowl should be made with 10 *palas* weight of copper, that the diameter at the opening should be 12 *aṅgulas* and the height 6 *aṅgulas.*[3] A highly ingenious mode of micro-measurement was adopted for the size of the perforation at the bottom. According to Āryabhaṭa, the hole should be such that a gold wire, 1 *pala* in weight and 8 *aṅgulas* long, should just pass through it.[4]

1.3 In Sanskrit, the bowl is called *ghaṭī* or *ghaṭikā* and these terms denote also the time unit measured by this device, viz. 24 minutes. The whole apparatus was accordingly called *Ghaṭīyantra* or *Ghaṭikāyantra*.

2.1 Outside the realm of astronomy / astrology, this sexagesimal division of the day-and-night into 60 *ghaṭīs* (and again of each *ghaṭī* into 60 *vighaṭīs* or *palas*) did not have much practical use in the pre-industrial society. For practical purposes, broader divisions sufficed. The *Arthaśāstra*, it may be recalled, divides the king's day into eight periods and likewise the night.[5] More common was the practice of dividing the daytime into four quarters styled *praharas* and the night into four quarters called *yāmas*.

2.2 In order to reconcile these two systems so that one *prahara/yāma* consisted of integral number of *ghaṭīs*, the duration of the *ghaṭī* was slightly reduced from 24 minutes to 22½ minutes. Thus the mean daytime (i.e. 6 am to 6 pm) was divided into 4 *praharas*, and each *prahara* into eight *ghaṭīs* (of 22½ minutes). According to I-Tsing, this was the custom prevailing in the Buddhist monasteries at Nalanda and Kuśinagara in the seventh century. (At Nalanda, however, the *prahara* was divided into 4 parts only).[6]

2.3 From the seventh century onwards, institutions for time-keeping are attested at royal palaces, Buddhist monasteries, town squares, etc. Here time was measured constantly with the *Ghaṭī-yantra*, and the passage of each *ghaṭī* and the completion of

3. Kripa Shankar Shukla, "Āryabhaṭa I's Astronomy with Midnight Day-reckoning", *Gaṇita* 18.1 (June 1967), p. 95, vv. 29-30.
4. Ibid.
5. *Arthaśāstra*, 1.19.6.
6. I-Tsing, *A Record of the Buddhist Religion as Practised in India and the Malay Archipelago (A.D. 671-695)*, tr. J Takakusu, Delhi 1966, pp. 144-146.

each *prahara/yāma* was announced through appropriate number of strokes on the drum, bell, or gong.

3.1 The institution was so popular that the Muslim rulers adopted not only the device but also the prevailing time units like *ghaṭī* and *pala*. Thus Fīrūz Shāh Tughluq set up a *ṭās-i ghariyāla* (lit. bowl and gong) at his palace gate which, in the words of his chronicler, aroused the wonder of people from Khurasan to Bengal.[7] In his memoirs, Bābur reports of the existence of a class of persons called *ghariyālīs*, whose profession it was to keep time by means of the gong in every major town of Hindustan. Bābur also made an improvement in the mode of announcing the *prahara* and *ghaṭī*.[8] The *Ā'īn-i Akbarī* also contains a long account on the institution of time-keeping. Though faulty in technical details, it informs that time keeping and announcing was royal prerogative under Akbar.[9] In the subsequent centuries, minor nobles also began to maintain water clocks at their gates and to cause time to be announced regularly—a practice that still survives at police stations and jails, though time is no more measured with water clocks.

3.2 That the institution of time keeping with water clocks was very popular in the Telugu country is attested by a number of references in literary works and inscriptions of the 13th-16th centuries.[10] Two inscriptions, in particular, record private endowments for maintaining water clocks in certain temples, so that the

7. Shams-i-Sirāj 'Afīf, *Tārīkh-i Fīrūz Shāhī*, ed. Maulwi Wilayat Hussain, Asiatic Society of Bengal, Calcutta, 1890, chapter 18, pp. 254-260.
8. *Bābur-Nāma*, tr. A.S. Beveridge, Delhi 1979, pp. 516-17.
9. Abū 'l-Fazl, *Ā'īn-i Akbarī*, ed. H. Blochman, Asiatic Society of Bengal, Calcutta, 1876, vol. I, pp. 9-10. On two miniature paintings from Akbar's atelier depicting the water clock, see Sreeramula Rajeswara Sarma, "Astronomical Instruments in Mughal Miniatures," *Studien zur Indologie und Iranistik*, 16-17 (1992), pp. 235-276; reprinted in this Volume, pp. 76-121.
10. Cf. M. Somasekhara Sarma, *History of the Reddi Kingdoms (circa 1325 A. D. to circa 1448 A. D.)*, Waltair 1948, pp. 324-327; M. Krishna Kumari, "Evidence of the Use of *Ghadiyaram* in Medieval Andhra," read at a seminar on "Kakatiyas and their Contribution to Art and Literature," Waltair, 1991.

rituals are performed according to a precise time schedule.[11]

4.1 Some more details abut this institution are available from the account of a Dutch traveller, Daniel Havart. In his *Open Ondergang van Coromandel*, he reports about a novel method of announcing time which he observed in the year 1676 at Hayatnagar.[12] His book was published in 1693. Since he had left India in March 1685 and did not come again, we accept his descriptions referring to the period of his stay in India, i.e. 1672-1685. This place is not far from Golconda and seems to have served as an outpost to the Golconda Fort. According to Tavernier, it lay on the route from Golconda to the port town Machilipatnam and had a large caravanserai with suits of rooms for travelling nobility as well.[13]

4.2 Daniel Havart states that the most conspicuous part of the village Hayatnagar was the gateway where time was measured with the water clock and announced in two different ways. First the passage of each *ghaṭī* and of each *prahara* was announced audibly through strokes on a copper basin or gong. Besides this common method, there was also a visual mode of announcing time. This was done by setting up on the gateway a certain number of small wooden planks, some painted red and some white. Red planks indicated the number of *praharas* elapsed since the mean sunrise at 6:00 am, and white planks indicated the number of *ghaṭīs* elapsed within the said *prahara*.

4.3 Havart illustrates the system by providing an example of how time is announced at 10:30 am. At this time, four and half hours have elapsed since the mean sunrise. Expressed in Indian time units, this duration equals to 1 *prahara* (= 3 hrs) and 4 *ghaṭīs* (4 times 22.5 = 90 minutes).[14]

11. The Sarpavaram inscription of 1404 AD (SII, vol. V. no. 8) records a private endowment for a water clock in the temple of Bhāvanārāyaṇa, while Tallapalli grant (*Reddi Sanchika*, ed. Vaddadi Appa Rao, Appendix) mentions the temple of Gopīnātha. Cf. Krishna Kumari, op. cit.
12. See the Appendix.
13. Jean Baptiste Travernier, *Travels in India*, tr. V. Ball, ed. W. Crooke, I, pp. 139-140; Irfan Habib, *An Atlas of the Mughal Empire*, New Delhi 1984, p. 60, middle column. Today, Hayatnagar is the outermost suburb of Hyderabad on the Vijayawada road.
14. Or 12 immersions of the bowl.

4.4 Now this time is announced in two different ways. First audible method: upon a copper basin, they strike four strokes for the passage of 4 *ghaṭīs*, and after a short pause one more stroke to indicate the passage of 1 *prahara* since sunrise. The pause is necessary in order to distinguish the two kinds of enumeration, viz., strokes for the *ghaṭī* and those for the *prahara*.[15] Furthermore, they also announced time through visual means, by setting up one red plank and four white planks upon the gateway, indicating the passage of 1 *prahara* since sunrise and 4 *ghaṭīs* thereafter.

4.5 Havart goes on to say "in the night it is likewise," which would mean that both types of announcement were made at night also. While the strokes on the gong can be heard at night, in fact, more distinctly than in the daytime, coloured planks upon the gateway cannot be seen unless they were illuminated with torches. Havart does not say expressly that the planks were illuminated, but that should not have been impossible.

5.1 Announcing time through strokes on the drum or on the gong was prevalent at least from the time of Buddhaghoṣa in the fifth century, and indeed this practice governs the vocabulary of time-telling in almost all Indian languages, as e.g. when we say *tīn baj rahe hain*. But indicating time by visual means has not been recorded elsewhere. Therefore Havart's account of this practice deserves some attention.[16]

5.2 However, this account raises two questions. First, how were adjustments made for the accensional difference, i.e. for the varying

15. In Bābur's time, instead of a pause, a series of rapid strokes separated the *ghaṭī* from the *prahara* (*Bābur-Nama*, tr. A.S. Beveridge, pp. 516-517). A relic of this practice survives still at the railway stations in South India. The porter first makes a number of rapid strokes, on a piece of suspended rail, for attracting the attention of the passengers, and then beats either three distinct strokes to announce the imminent arrival of the up train to Madras, or two distinct strokes to indicate the arrival of the down train from Madras.

16. The Dutch are said to have presented the Shah of Persia sinking bowl water clocks (*gory schotels*), which they had taken from India, cf. Cornelis Spelman, *Journal der reiz von den gezant der O. I. Compagnie Joan Cunaeus naar Iarzie in 1651-1652,* 19-8, p. 146, cited in O. Kurz, *European Clocks and Watches in the Near East*, London 1975, p. 4, n. 5.

length of the daylight hours. Havart is silent on this question.[17] The second question is as follows: Does announcing time visually have any advantage over gong strokes? Gong strokes, it may be noted, are made after every 22½ minutes. If a traveler arrives at the gate of Hayatnagar, say a few minutes after 10:30, he would miss the last announcement and has to wait at least 20 minutes to know what the time of the day is. But if there are visible planks, he will know time even if he arrives in the middle of a particular *ghaṭī*. But then, in the pre-industrial society of 1676, did it matter much whether the time was 10:30 or 10:52,30 ? Be that as it may, it is interesting to note that the Dutch factors thought it worthwhile to emulate this practice of keeping and announcing time. But, instead of the water clock[18] of 22½ minutes duration, they employed sand glasses of half an hour's duration.

APPENDIX

Daniel Havart, M.D., *Open Ondergang van Coromandel*, Amsterdam 1693, II, pp. 205-206.

"The [main] landmark of the village Hayatnagar . . . is a gateway . . . and above that gate one sees small red and white wooden planks [which indicated] how many hours of the day have [already] passed . . .

"Muslims, who do not have mechanical clocks (*slaande klokken*), count the hours of the day and of night in the following manner. They divide the day into four parts, each of three hours: because [the measurement of] the day always begins at six in the morning and terminates at six in the evening: each part of the day being once again subdivided into eight *gharis (gerrijis)*, and each *ghari* is counted by means of (lit. according to) a certain small cup (*kopja*) with a hole [at the bottom] which, [when] placed upon water, comes to sink in one and half quarter hours' time, and then one

17. John Gilchrist, "Account of the Hindustanee Horometry," *Asiatick Researches*, 5 (1795), pp. 81–89, discusses some methods of adjustment as practiced in Bengal.
18. Considering their universal employment, there ought to be scores of water clocks preserved in India and abroad. However, only about half a dozen pieces are known in foreign collections and about an equal number in India, one of which is prominently displayed in the gateway of the fort at Golconda.

ghari has ended, when they also strike one stroke upon a copper basin (*bekkan*).

"It is now, for example, half past ten, then they strike first four and then, after a brief pause (*perposing*), one [more] stroke, indicating that these four *gharis* and one quarter of the day have elapsed, and at the same time, they also set up there four white and one red planks at the top of the gateway, where everyone can always see what the time is. In the night it is likewise.

"The Dutch also do so, except that in the place of the small cup, they use hour glasses of half an hour [duration] and consequently have not more than six [half hours] in one quarter of the day."

ACKNOWLEDGEMENTS

Grateful thanks are due to Professor M. Krishna Kumari, Department of Archaeology, Andhra University, Waltair, for kindly providing us with a copy of her unpublished paper "Evidence of the Use of *Ghadiyaram* in Medieval Andhra".

7

Measuring Time with Long Syllables: Bhāskara I's Commentary on Āryabhaṭīya, Kālakriyāpāda 2

One of the most valuable documents on Indian astronomy published in recent times is Bhāskara I's commentary on the *Āryabhaṭīya*.[1] Unfortunately the commentary is not complete; it breaks off at *Golapāda* 6. In the available portion too there are occasionally readings which are not satisfactory. One such reading occurs in the commentary on the second verse of the *Kālakriyāpāda*. In this verse, Āryabhaṭa states that the time taken to utter sixty long syllables (*gurvakṣara*) is one *vināḍikā* of a sidereal day.[2]

First published in: *Indian Journal of History of Science*, 36 (2001) 51-54.

1. *Āryabhaṭīya of Āryabhaṭa* with the Commentary of Bhāskara I and Someśvara, critically edited with Introduction and Appendices, by Kripa Shankar Shukla, Indian National Science Academy, New Delhi 1976.
2. Āryabhaṭa is the first one to introduce a completely sexagesimal division of time in analogy to the sexagesimal division of the circle. In this scheme, one *nāḍī* or *nāḍikā* is divided into 60 *vināḍikās*, which are further subdivided into 60 *gurvakṣaras* each. Thus the *gurvakṣara* is the 3600th part of the *nāḍī* and equals 0.4 seconds. Some scholars erroneously equate this *gurvakṣara* with a time unit called *akṣara* defined in verse 39 of the Yajus recension of the *Vedāṅga Jyotiṣa*. Commenting on this verse, A. Weber (*Über den Vedakalender, namens Jyotisham,* Berlin 1862, p. 104, n. 3) opines that these *akṣaras* are the same as the long syllables of two *mātras*. T.S. Kuppanna Sastry shares the view (*Vedāṅga Jyotiṣa of Lagadha, in its Ṛk and Yajus Recensions*, with Translation and Notes by T.S. Kuppanna Sastry, critically edited by K.V. Sarma, Indian National Science Academy, New Delhi 1985, p. 38: "The *akṣara* mentioned here is the length of time called *gurvakṣara*, equal to two *mātras* of time"). However, it is obvious that

But then, if a man utters sixty long syllables very fast, it will be less than one *vināḍikā*; or if he utters them very slowly it can be more than one *vināḍikā*. Refuting such a possible objection, Bhāskara proclaims that the sixty syllables should be uttered neither too fast nor too slowly, but at a middling speed.

To this one may again object by saying: how can you say middling speed when Āryabhaṭa himself did not specify this. Bhāskara counters this by stating that in all cases where no specification is made, one should take the middle course (*loke anirdiṣṭeṣu kāryeṣu madhyamaprāptiḥ*). Having thus established the speed at which the long syllables should be uttered, Bhāskara adds that the following 60 long syllables constitute the duration of one *vināḍikā*:

"māsānte pakṣasyānte sa hy ākāśe deśe svaṃ miśraṃ vakraṃ kāntaṃ vṛttaṃ pūrṇaṃ candraṃ sattvād rātrau te kṣutkṣāma prādante śveto prājyo krūras tasmād vānte harmyasyāntaḥ saṃsuptasyaikānte kartavyā /"
etāni ṣaṣṭir gurvakṣarāṇi vināḍikākālah // [3]

These are indeed sixty long syllables. If one recites these at a middling speed, it should take one *vināḍikā*. But these syllables make no sense and appear to be badly garbled. The manuscripts used in the edition do not seem to offer any alternative readings of this passage. Neither Sūryadevayajvan[4] nor Parameśvara,[5] whose commentaries on the *Āryabhaṭīya* are available in print, cite this passage.

In these circumstances, it is fortunate to find the correct reading of this passage in an unpublished manuscript entitled *Ghaṭikāyan-*

---

these two units cannot have the same duration. According to the *Vedāṅga Jyotiṣa*, 1 *naḍikā* consists of 10 1/20 *kalās*, 1 *kalā* of 124 *kāṣṭhās*, and 1 *kāṣṭhā* of 5 *akṣaras*. Accordingly there will be 6231 *akṣaras* in 1 *nāḍikā* as against 3600 *gurvakṣaras* of Āryabhaṭa.

3. *Āryabhaṭīya of Āryabhaṭa* with the Commentary of Bhāskara I and Someśvara, p. 175.
4. *Āryabhaṭīya of Āryabhaṭa* with the Commentary of Sūryadeva Yajvan, critically edited with Introduction and Appendices, by K.V. Sarma, Indian National Science Academy, New Delhi 1976
5. *Āryabhaṭīya of Āryabhaṭa* with the Commentary *Bhaṭadīpikā* of Paramādīśvara [= Parameśvara], ed. H. Kern, Leiden 1874.

traghaṭanāvidhi ("Method of Setting up the Water Clock").[6] After explaining how to set up the water clock for determining the auspicious moment of the wedding, this anonymous work cites the correct version of the passage which got jumbled up in the manuscript tradition of Bhāskara I's commentary. In the *Ghaṭikāyantraghaṭanāvidhi*, the passage reads thus:

mā kānte pakṣasyānte paryākāśe deśe svāpsīḥ
kāntaṃ vaktraṃ vṛttaṃ pūrṇaṃ candraṃ matvā rātrau cet /
kṣutkṣāmaḥ prātaṃś cetaś ceto rāhuḥ krūraḥ prādyāt
tasmād dhvānte harmyasyānte śayyaikānte kartavyā //

"Do not, O pretty one, at the end of the [bright] fortnight, sleep at a place open to the sky. Should it turn night, the cruel Rāhu, starving with hunger and roaming hither and thither, may eat you up, taking your pretty round face for the full moon. Therefore, after darkness, make your bed at a secluded place inside the house."

It is a pretty poem, consisting of only long syllables. Evidently, this is a verse with four uniform feet (*sama-vṛtta*) and each foot consists of five *ma-gaṇas*. Indeed, this metre is noticed in several works on prosody, albeit with different nomenclature. Thus Kedāra's *Vṛttaratnākara* and Hemacandra's *Chando'nuśāsana* call this metre *Kāmakrīḍā*. *Chandomañjarī* of Gaṅgādāsa designates it as *Līlākhela*. It receives an unusual designation *Jyotiḥ* or *Mitra* in Jayakīrti's *Chando'nuśāsana*, and the anonymous *Prākṛtapaiṅgala* styles it *Sāraṅgī*.[7] But all these manuals on prosody belong to periods much later than Bhāskara I's own time.

A single recitation of the verse at an even pace will last one *vināḍika*, which is also known as *vighaṭikā* or *pala*, and which equals 24 seconds. It is the one-sixtieth part of the *nāḍikā*, which is also called *nāḍī*, *ghaṭikā* or *ghaṭī*. This is equal to 24 minutes and it was the standard unit of time measurement in pre-modern India.

The *Ghaṭikāyantraghaṭanāvidhi* suggests that this and similar verses in this metre are recited after installing the water clock

6. MS No. 37074, Sarasvati Bhavan Library, Sampurnananda Samskrita Vishvavidyalaya, Varanasi.
7. H.D. Velankar (ed), *Jayadāman (A Collection of Ancient Texts on Sanskrit Prosody and a Classified List of Sanskrit Metres with an Alphabetical Index)*, Bombay 1949, p. 135.

(*Ghaṭikāyantra*). A story in the *Kathāratnākara,* composed by Hemavijaya Gaṇī at Ahmedabad in 1600 AD,[8] informs that such verses are called *pala-vṛttas* (i.e. verses for measuring a *pala*).

Though there is no clear statement in any text, it is possible that the *pala*-verses were employed in measuring time in the following manner. A hemispherical copper bowl (*ghaṭikā*) with a fine hole at its bottom is set up on the surface of the water in a larger receptacle (*kuṇḍa*). Each immersion of the bowl measures one *ghaṭikā.*[9] For measuring the sub-multiples of *ghaṭikā,* the height of the bowl can be sub-divided with appropriate markings. To do so geometrically is difficult, but one can empirically divide the bowl, say into six or ten parts. Each part then would denote either 10 *palas* (= 4 minutes) or 6 *palas* (= 2 minutes and 24 seconds). Gilchrist reports about the existence of water clocks with such sub-divisions towards the close of the eighteenth century.[10] But in the course of my survey of Indian time measuring instruments, I have not come across even a single specimen of such a bowl. All the bowls I have seen are without sub-divisions. With these one can measure only full *ghaṭikās.* How does one then measure fractions? One possible method is to measure the complete *ghaṭikās* by means of the water clock and the fractions in terms of *palas* by reciting the *pala*-verses. Thus, if an auspicious moment is set for 10 *ghaṭikās* and 15 *palas* after sunrise, the water clock is set up exactly at sunrise. After ten immersions of the bowl are complete, the *pala* verse is recited 15 times. Then the time will be 10 *ghaṭikās* 15 *palas* after sunrise.

8. Hemavijayagaṇī, *Kathāratnākara*, published by Hiralal Hansaraj, Jamnagar, 1911, No. 211, pp. 538-540. Translated into German by Johannes Hertel, *Kathāratnākara: Das Märchenmeer: Eine Sammlung indischer Erzählungen*, 3 vols, Munich 1920, pp. 253-255.
9. On measuring time with water clocks, see S.R. Sarma, "The Bowl that Sinks and Tells Time," *India Magazine, of her People and Culture,* 14.9 (September 1994) 31-36; reprinted in this Volume, pp. 125-135. See also Virendra Nath Sharma, "Astronomical Instruments at Kota," *Indian Journal of History of Science*, 35 (2000) 233-244, esp. Figs. 4 and 5.
10. John Gilchrist, "Account of the Hindustanee Horometry," *Asiatick Researches*, 5 (1795) 81-89, esp. 87: "These *kutorees* [bowls] are now and then found with requisite divisions, and subdivisions, very scientifically marked in *Sanscrit* characters. . . ."

8

Setting up the Water Clock for Telling the Time of Marriage

0.1 From about the fourth century AD up to recent times the water clock of the sinking bowl type (*Ghaṭikā-* or *Ghaṭī-yantra*) has been the chief device in India for measuring time.[1] The instrument consists of a hemispherical bowl (*ghaṭikā* or *ghaṭī*) with a minute perforation at the bottom. When this bowl is placed on the surface of water in a larger vessel or basin (*kuṇḍa, kuṇḍikā, kuṇḍī*), water slowly percolates into the bowl through the perforation. When the bowl is full, it sinks to the bottom of the basin. The weight of the vessel and the size of the perforation are so regulated that the bowl sinks sixty times in a nychthemeron (*ahorātra*). Thus the time taken for filling the bowl once is one-sixtieth part of a nychthemeron, or twenty-four minutes. This was the standard unit of time measurement in India and is called *ghaṭikā* or *ghaṭī* after the name of the bowl. The *ghaṭikā* is subdivided into sixty *vighaṭikās*, which are also called *palas*.[2]

When the bowl sinks to the bottom of the basin, indicating the completion of one *ghaṭikā*, this fact is broadcast with blasts on a conch-shell or strokes on a drum. In the early medieval period, the

First published in: Chales Burnet et al (eds), *Studies in the History of Exact Sciences in Honour of David Pingree*, Brill, Leiden, 2004, pp. 302-30.

1. On the different types of water clocks, see Needham, p. 115; on the sinking bowl type in India, see S.R. Sarma, 'The Bowl that sinks and tells Time'; reprinted in this Volume, pp. 125-135.
2. This sinking bowl type replaced the outflow type which too measured the standard unit of twenty-four minutes. Here the instrument and also the time unit were styled *nālikā, nālī* or *nāḍikā, nāḍī*. Even after the outflow clock was discarded, the old names continued to be used along with *ghaṭikā* and *ghaṭī*.

conch and drum were replaced by the gong, which was designated in the Middle Indic as *ghaḍiyāla* (from *ghaṭikālaya*, 'water clock house').

0.2 Āryabhaṭa I was the first to describe the bowl of this water clock in his *Āryabhaṭasiddhānta.*[3] He describes it in the following words:

vṛttaṃ tāmramayaṃ pātraṃ kārayed daśabhiḥ palaiḥ /
ṣaḍaṅgulaṃ tadutsedho vistāro dvādaśānane //
tasyādhaḥ kārayec chidraṃ palenāṣṭāṅgulena tu /
ity etad ghaṭikāyantraṃ palaṣaṣṭyāmbupūraṇāt //
sveṣṭaṃ vānyad ahorātre ṣaṣṭyāmbhasi nimajjati /
tāmrapātram adhaśchidram ambuyantraṃ kapālakam //

Kripa Shankar Shukla's translation of this passage reads thus:

'One should get a hemispherical bowl made of copper, 10 *palas* in weight, six *aṅgulas* in height, and twelve *aṅgulas* in diameter at the top. At the bottom thereof, let a hole be made by a needle eight *aṅgulas* in length and 1 *pala* in weight.
'This is the *ghaṭikā-yantra*, (so named) because it is filled by water in a period of 60 *palas* (i.e. one *ghaṭī*).
'Any copper vessel made according to one's liking with a hole in the bottom, which sinks into water 60 times in a day and night, is the water instrument called *Kapāla.*'

Shukla obviously thought that this passage describes two different instruments, called respectively *Ghaṭikā-yantra* and *Kapāla-yantra.*[4] But the fact is that these are two different methods of producing one and the same *Ghaṭikā-yantra*. The first method is to take ten *palas'* weight of copper and with that to produce a hemispherical bowl, having a height of six *aṅgulas* and the upper diameter of twelve *aṅgulas*. There should be a fine perforation at the bottom of the bowl. The small size of the perforation is defined in a peculiar

3. This work is now lost, but luckily Āryabhaṭa's descriptions of various instruments survive in Rāmakṛṣṇa Ārādhya's commentary on the *Sūryasiddhānta.* Kripa Shankar Shukla published these passages in his paper 'Āryabhaṭa I's Astronomy with Midnight Day-Reckoning'.
4. Accordingly, *Indian Astronomy: A Source-Book* reproduces these verses and their translation under two different heads, viz. 10.11.1: *Ghaṭikā-yantra* (p. 91) and 10.5.5: *Kapāla-yantra* (p. 89).

manner. It should be made, as Āryabhaṭa says, 'with [an object, say a needle, having a weight of] one *pala* and a [length of] eight *aṅgulas*.' He does not specify what kind of metal is to be used for the needle. But since the weight and the length are prescribed, the implication is that a wire of uniform thickness and a specific length has to be produced from a given amount of the metal. Such metal could only be gold; it is malleable enough to be drawn into wires of uniform thickness. Indeed, in his *Arthaśāstra*, Kauṭilya prescribes that the outflow water clock (*nālikā*) should have a 'perforation by [a needle made of] four *māṣakas* of gold and four *aṅgulas* in length.'[5] It is highly probable that the unspecified metal in Āryabhaṭa's verse is also gold. But a thin gold needle cannot pierce through a copper bowl. Hence Āryabhaṭa's specification (and also Kauṭilya's) should be understood to mean that 'the perforation should be such that a gold wire, one *pala* in weight and eight *aṅgulas* in length, can pass through it.'[6]

Goldsmiths of that time may have been able to draw fine grades of gold wire, but whether they could draw a wire measuring exactly eight *aṅgulas* from a lump of gold weighing exactly one *pala* is open to question. However, the main purpose of the instrument is to measure one-sixtieth part of the nychthemeron. For this purpose, the bowl should be such that it fills with water and sinks into the

5. This system is employed for the first time in the *Kauṭilīya Arthaśāstra* for defining the perforation of the outflow water clock (*Nālikā-* or *Nāḍikā-yantra*); cf. 2.20.34: *suvarṇamāṣakāś catvāraś caturaṅgulā-yāmāḥ kumbhacchidram āḍhakam ambhaso vā nālikā* //
6. Similar specifications were given in other texts as well. The *Jyotiṣ-karaṇḍaka* (verses 11-14) lays down that the hole of the outflow water clock (*nāligā*) should be such that ninety-six hairs from the tail of a three years' old female elephant calf, or twice that number from the tail of a two years' old female elephant calf, or a gold needle of four *māṣas*' weight and four *aṅgulas*' length should pass through it. Al-Bīrūnī (vol. I, p. 334) cites from 'the book *Srûdhava* by Utpala the Kashmîrian': 'If you bore in a piece of wood a cylindrical hole of twelve fingers' diameter and six fingers' height, it contains three *manā* of water. If you bore in the bottom of this hole another hole as large as six plaited hairs of a young woman, not of an old one nor of a child, the three *manā* of water will flow out through this hole in one *ghaṭī*.' It may be noted that this is also an outflow water clock.

basin sixty times in a day and night; in other words, the bowl should fill in a period of one *ghaṭikā* or of sixty *palas*. Therefore, Āryabhaṭa goes on to suggest that the measurements of the bowl and of the needle are not so important; what is important is that the bowl should be able to fill with water sixty times in a day and night. This is the import of verse no. 31, where particle '*vā*' indicates that this is an alternative method. The word '*kapāla*' here is not the designation of another instrument, but denotes the hemispherical shape of the bowl. In this light, I retranslate the verse:

Or, alternatively, any hemispherical (*kapāla*) copper vessel made according to one's liking with a perforation in the bottom, which sinks into water sixty times in a day and night, is the water clock (*ambu-yantra*).[7]

0.3 Lalla, while retaining the dimensions of the bowl, changed the size of the perforation:

daśabhiḥ śulbasya palaiḥ pātraṃ kalaśārdhasannibhaṃ ghaṭitam /
hastārdhamukhavyāsaṃ samaghaṭavṛttaṃ dalocchrāyam //
satryaṃśamāṣakatrayakṛtanalayā samasavṛttayā hemnaḥ /
caturaṅgulayā viddhaṃ majjati vimale jale nāḍyā // [8]

The bowl, which resembles half a pot (i.e. hemispherical), which is made of ten *palas* of copper, which is half a cubit (i.e. twelve *aṅgulas*) in diameter at the mouth and half (i.e. six *aṅgulas*) as high, which is evenly circular, and which is bored by a uniformly circular needle, made of three and one-third *māṣas* of gold and of four *aṅgulas* in length, sinks into clear water in one *ghaṭikā* (*nāḍī*).

It is difficult to say what caused this shift in the size of the

7. Brahmagupta too does not prescribe any measurements for the bowl and the perforation. All that is required is that the bowl should sink sixty times in an *ahorātra*; cf. *Brāhmasphuṭa-siddhānta*, 22.41:
ghaṭikā kalaśārdhākṛti tāmraṃ pātraṃ tale 'pṛthu cchidram /
madhye tajjalamajjanaṣaṣṭyā dyuniśaṃ yathā bhavati //
'The *Ghaṭikā-yantra* is a copper vessel of the shape of a hemisphere. At the centre of the bottom is a small perforation so made that the bowl sinks sixty times in a day and night.' *Sūryasiddhānta*, 13.23 also gives a similar definition, without specifying the measurements:
tāmrapātram adhaśchidraṃ nyastaṃ kuṇḍe 'malāmbhasi /
ṣaṣṭir majjaty ahorātre sphuṭaṃ yantraṃ kapālakam //
8. *Śiṣyadhīvṛddhidatantra*, part I, 21 (*Yantrādhikāra*) 34-35.

perforation—from the gold needle of one *pala* weight and eight *aṅgulas'* length to a gold needle three and one-third *māṣas* in weight and four *aṅgulas* in length.[9] According to Śrīdhara, who is a near contemporary of Lalla, sixty-four *maṣas* make one *pala*.[10] Then the perforation prescribed by Lalla would be 5/48th of that prescribed by Āryabhaṭa.[11] While the size and weight of the bowl remained the same, such stark reduction in the size of the perforation would have greatly increased the duration of the time needed for the bowl to become full and then to sink, and consequently the duration of the *ghaṭikā* as well. One is therefore led to suspect that these specifications for the size of the perforation in terms of a gold wire of certain weight and length are fictitious and have no connection with actual practice. Yet this latter specification for the size of the perforation is repeated in several subsequent works.

0.4 Bhāskara II dismisses these specifications of the size as impractical. According to him a bowl of any size with any arbitrary hole can serve as the water clock. It need not even have a duration of one *ghaṭikā*; a bowl of any duration will do for measuring time. Thus he declares in the *Siddhāntaśiromaṇi*:

9. Of course, Lalla also adds that the vessel and the perforation can be of any dimension provided the bowl sinks in one *ghaṭī*; cf. *Śiṣyadhīvṛddhidatantra*, part I, 21 (*Yantrādhikāra*), 36:

   athavā svecchāghaṭitaṃ ghaṭīprmābhiḥ prasādhitaṃ bhūyaḥ /
   'Or, it is a vessel manufactured according to one's liking [with a perforation at the bottom] which has been adjusted by the measure of a *ghaṭī* by repeated [trials].' (Bina Chatterjee's translation, slightly modified).
10. *Pāṭīgaṇita*, Rule 10.
11. Let A be the area of the cross-section of the wire (and therefore the area of the hole produced by the wire) of 1 *pala* weight and 8 *aṅgulas'* length. Then its volume will be $8A$. If a gold wire of 1 *pala* (64 *māṣas*) has a volume of $8A$, then a gold wire of $3\frac{1}{3}$ *māṣas* will have the volume of $\frac{8A}{64}\cdot\frac{10}{3}=A\cdot\frac{5}{12}$. The gold wire of $\frac{10}{3}$ *māṣas* is 4 *aṅgulas* long. Then the area of its cross-section $B=A\cdot\frac{5}{12}\cdot\frac{1}{4}=A\cdot\frac{5}{48}$.

ghaṭadalarūpā ghaṭitā ghaṭikā tāmrī tale 'pṛthucchidrā /
dyuniśanimajjanamityā bhaktaṃ lyuniśaṃ ghaṭīmānam //[12]

A copper bowl, formed like a hemisphere, having a small (*apṛthu*) hole at the bottom. The duration of a day and night divided by the number of immersions [of this bowl] gives the measure of the water clock.

This statement is elaborated in his auto-commentary *Vāsanā-bhāṣya*: 'Here, we ignore the definition of the water clock given by certain scholars as '*daśabhiḥ śulbasya palair . . .*' because it is illogical (*yukti-śūnya*) and difficult to implement (*durghaṭa*).'[13]

Here Bhāskara is attacking Lalla directly by quoting the latter's definition of the water clock. In his view, such specifications for the size of the bowl and of the perforation are impractical. On the other hand, by stating that a bowl with any duration of immersion will do, Bhāskara goes to the other extreme. If the bowl sinks, say, in thirty-four minutes and thus measures a period of thirty-four minutes each time it sinks, these periods have to be converted into *ghaṭikās* of twenty-four minutes. By not insisting that the bowl should sink in one *ghaṭikā*, as Āryabhaṭa and Brahmagupta did, Bhāskara underestimates the Indian artisan's ability to fabricate bowls of exactly one *ghaṭikā* duration.

0.5 There is yet another problem with Bhāskara II's statement. His diction (*ghaṭadalarūpā ghaṭitā ghaṭikā tāmrī tale 'pṛthucchidrā*) follows very closely that of Brahmagupta (*ghaṭikā kalaśārdhākṛti tāmraṃ pātraṃ tale 'pṛthu cchidram*).[14] Since the manuscripts rarely mark the *avaghraha* symbols, these have to be inserted according to the context by the modern readers. There is no way of knowing whether the author himself desires an *avagraha* at a certain point or not, unless this is spelt out clearly in a subsequent statement by the author himself, or in an auto-commentary or in a reliable commentary written by one who is close to the author in time or spirit. In the present case, it is uncertain whether Brahmagupta and Bhāskara II consider the perforation to be '*pṛthu*' or '*apṛthu*'. Bhāskara's auto-commentary is silent on this issue.

12. *Siddhāntaśiromaṇi, Golādhyāya, Yantrādhyāya*, 8, pp. 366-367.
13. Ibid., p. 367: *atra daśabhiḥ śulbasya palair ity ādi yad ghaṭīlakṣaṇaṃ kaiścit kṛtaṃ tad yuktiśūnyaṃ durghaṭaṃ cety etad upekṣitam /*
14. See n. 6 above.

However, Bhāskara II's commentator Munīśvara (b. 1603) expressly prohibits the insertion of an *avagraha* here. Thus he argues in his *Marīci* commentary: 'The bowl should be so made that it has a large hole (*pṛthucchidra* = *mahārandhra*) at the bottom. Through this statement it is indicated that the hole should be made in such a manner that, when the bowl is placed on the water of the basin and when water enters [the bowl], the hole is not blocked by any dirt that may be in the water of the basin. Because of the possibility of a small hole getting blocked by dirt and the like, assuming here a coalescence of the vowel *a* (*akāra-praśleṣa*) [by reading *apṛthu*] is not proper.'[15]

We do not know if this is the intention of Bhāskara II or if this is just Munīśvara's view.[16] But a large hole will certainly make the bowl sink very rapidly. If it sinks, for example, in ten minutes, then it has to be lifted, emptied and placed again on the surface of the water each ten minutes! There is no practical advantage in using a water clock of such short duration.[17] It is more likely, therefore, that Bhāskara intended to say '*apṛthu*'. All the extant specimens that I have seen have very small holes.

0.6 Besides the question of the size of the perforation, another misunderstanding seems to have arisen about the volume of the bowl. The basic requirement, as we saw, is that the bowl should sink sixty times in a day and night, or that it should become full of water in one *ghaṭī* of sixty *palas*. Āryabhaṭa expressed this as '*palaṣaṣṭyā ambupūraṇa*'. This seems to have been misunderstood to mean that the bowl should have the capacity to hold sixty *palas* of water. In his commentary on the *Āryabhaṭīya*, Bhāskara I mentions such a view, albeit as a *pūrvapakṣa*.[18] Since his discussion

15. *Siddhāntaśiromaṇi*, p. 367: *tale 'dhobhage pṛthucchidrā mahārandhrā ghaṭikā kāryā / etena jalapūrṇapātre nikṣiptādhaśchidre jalāgamane jalāntargatamalādikaṃ vastu pratibandhakaṃ na bhavati tathā chidraṃ kāryam iti sūcitam / sūkṣmacchidre malādikena tatpratibandha-sambhavād akārapraśleṣo na yuktaḥ /*
16. The only other published commentary by Nṛsiṃha is silent on this issue; see p. 442 of Nṛsiṃha's commentary. See Bibliography.
17. For preventing the blockage of the hole, it is generally prescribed that the water should be very clean.
18. *Āryabhaṭīya*, p. 174.

throws up many problems, it is necessary to quote his statement in full and then to analyse it:

kathaṃ punar divasasya ṣaṣṭibhāgaḥ sādhyate ity atrābhidhīyate / atra kecid bruvate suvarṇarajatatāmrāṇām anyatamaṃ pātram ardhavṛttākāraṃ ṣaṣṭipalapānīyadhārakaṃ pūrakaṃ nisrāvakaṃ vā ghaṭiketi / naiṣa niyamo[19] *yāvat palāni ṣaṣṭiḥ pānīyaṃ prasravaty āpūryate vā tāvatā nāḍikākāla iti / prājñās tu naivam iti mayante / kathaṃ tarhi / ahorātraprasrutasya pānīyasya ṣaṣṭibhāgo ghaṭikāpramāṇa iti sthūlaḥ kalpaḥ / sūkṣmas tu samāyām avanau nirdiṣṭākārasya śaṅkor ghaṭikāchāyām aṅkayitvā ghaṭikā sādhyate / ghaṭikāchidraṃ ca chāyākālavaśād yuktyā yojayitavyam /*

'How then is the one-sixtieth part of a nychthemeron to be determined?' To this question, [the following] has to be said. In this connection some say: 'The *Ghaṭikā*[*-yantra*] is a vessel [made out] of one of the metals like gold, silver or copper, hemispherical in shape (lit. semicircular), which holds (*dhāraka*) sixty *palas* of water and which is filled with or discharges [the same amount of water].' Actually there is no such rule that the duration of the *ghaṭikā* (*nāḍika-kāla*) is so long as it takes the vessel to discharge or to fill with sixty *palas* of water. Wise persons think that this is not so. How then?
It is only a rough method (*sthūlaḥ kalpaḥ*) to say that the one-sixtieth part of the water that has been discharged in the course of a nychthemeron is the measure of one *ghaṭikā*. The more accurate method is to measure the *ghaṭikā* by marking the shadow of one *ghaṭikā*, cast by a gnomon of specified shape that has been set up on a level ground. The perforation in [the bowl of] the *Ghaṭikā-yantra* should be made skilfully according to the period measured by the shadow.

Though Bhāskara I does not himself approve of it, there are some who hold that the *Ghaṭikā-yantra* should be a hemispherical vessel, which has the capacity to hold sixty *palas* of water and which either discharges or is filled with the same amount of water. Thus according to these persons both the outflow type of water clock which discharges water and also the sinking bowl type which is filled with water should have the same hemispherical shape and a volume of sixty *palas*. It is not known who these persons were who thought (already before AD 629 when Bhāskara wrote the

19. The Edition reads *naiṣa niyamaḥ* and closes the sentence here. I combine this sentence with the next one in order to draw a coherent meaning.

commentary) that the bowl should have the capacity of sixty *palas*, but this notion continued to be held in later times as well.

Thus Śrīpati in his *Siddhāntaśekhara* prescribes that the bowl should hold sixty *palas* of water:

śulbasya digbhir vihitaṃ palair yat
ṣaḍaṅguloccaṃ dviguṇāyatāsyam /
tad ambhasā ṣaṣṭipalaiḥ prapūryaṃ
pātraṃ ghaṭārdhapramitaṃ ghaṭī syāt //
satryaṃśamāṣatrayanirmitā yā
hemnaḥ śalākā caturaṅgulā syāt /
viddhaṃ tayā prāktanam atra pātraṃ
prapūryate nāḍikayāmbunā yat // [20]

A vessel, resembling half a pot in shape (i.e. hemispherical), made of ten *palas* of copper, six *aṅgulas* in height and twice the same in the diameter of the mouth, which can be filled with sixty *palas* of water, is the *Ghaṭī-yantra*. It should be pierced beforehand by a four *aṅgulas* long gold needle that has been made of three and one-third *māṣas* [of gold]. Then it fills with water [and sinks] in one *ghaṭikā* (= *nāḍikā*).

The expression '*tad ambhasā ṣaṣṭipalaiḥ prapūryaṃ*' can also be interpreted to mean that 'it should fill with water in a period of sixty *palas*' but then it would be redundant to say '*prapūryate nāḍikayāmbunā yat*'. Therefore, the first expression should pertain to the volume of the bowl. If the bowl really contains sixty *palas* of water, one *pala* would be 7.54 cubic *aṅgulas*. *Pala* occurs as a unit of weight and also as a unit of time; but this is a rare use of *pala* as a unit of liquid measure. But the chances are that this too is a fictitious specification for the capacity of the bowl.

On the other hand, the time unit *pala* is occasionally referred to as '*pānīyapala*' in contexts where it is quite certain that it is a time unit. Thus Āryabhaṭa II speaks of six *asus* in one *pānīyapala*; and Bhāskara II of 3600 *pānīyapalas* in a day and night.[21] Probably here it means 'a *pala* that is measured by means of water clock' as distinct from '*pala*', the unit of weight.

20. *Siddhāntaśekhara* 19.19-20.
21. See *Mahāsiddhānta*, 1.6: *prāṇāḥ pānīyapale tā; Siddhāntaśiromaṇi, Golādhyāya, Yantrādhyāya*, 8, *Vāsanābhāṣya: dyuniśanimajjana-saṃkhyayā yadi ṣaṭtriṃśacchatāni pānīyapalāni labhyante tadaikena kim iti trairāśikam.*

0.7 A third problem is the occasional confusion between two types of water clocks, namely the sinking bowl type and the outflow type that preceded it chronologically. While the former fills itself with water through a perforation and sinks in a fixed duration of time, the latter empties itself of water through a perforation in a fixed duration of time. In the previous paragraph we have seen that the *pūrvapakṣins* cited by Bhāskara I treat the vessels in both the types as hemispherical with the same volume. Or is it Bhāskara I himself who sees both the vessels as identical in shape and size? While the bowl in the sinking bowl type is certainly hemispherical or nearly so, this cannot be true of the outflow type, where the vessel must have had the shape of a regular cylinder or that of a truncated cone. Such confusion between the two types occurs elsewhere also. As we shall see below in 3.4.1-2, the *Dharmasindhu*, while discussing the sinking bowl type of water clock, cites in support a passage from the *Bhāgavata*, which describes the outflow type.

Sanskrit astronomical texts describe a large number of instruments. But no other instrument received such detailed specifications as the *Ghaṭikā-yantra* did. Again, of all the instruments described in the texts, the *Ghaṭikā-yantra* was the only one which was manufactured in great numbers and was used by astronomers and laymen alike. It is indeed strange that there should be such confusion in describing this simple instrument in all the highly 'scientific' texts on astronomy.

0.8 In spite of this theoretical confusion in the texts, countless specimens of this water clock were produced throughout the centuries and these kept reasonably correct time of one *ghaṭikā* of twenty-four minutes. But it is doubtful if any artisan has ever produced a bowl according to the textual prescriptions. In the few specimens that survive in modern collections,[22] rarely any bowl has the exact shape of a hemisphere; some are more conical, some are shallower, than a precise hemisphere. The sizes and weights too do not conform to the textual prescriptions, and vary considerably.

22. While cataloguing the extant specimens of Indian astronomical and time measuring instruments, I had occasion to study several water clocks.

The holes were obviously made by a trial and error method, by comparing the new bowl with another bowl that shows correct time or with a sundial, as Bhāskara I recommends,[23] and by suitably enlarging the hole or by reducing its size rather than by means of a gold wire of given dimensions. Sometimes the hole can become larger by constant use. Then the size of the hole can be reduced by hammering the area around the hole. In one case, a gold buff was added to the hole in order to reduce its size.[24]

0.9 There are evidences of time-keeping establishments in royal palaces, Buddhist monasteries, temple courtyards or town squares, maintained by royal or private endowments. Here time was constantly measured by this water clock and broadcast by conch and drum, or by the gong.[25] The Chinese traveller I-Tsing who spent some ten years from *c.* AD 675 to 685 at the famous Buddhist monastery of Nalanda, gives a detailed description of the time-keeping establishment there.[26] At the beginning of the eleventh century, al-Bīrūnī describes the time-keeping establishment at Purshor (modern Peshawar) and adds that 'Pious people have bequeathed for these clepsydrae (i.e. water clocks), and for their administration, legacies and fixed incomes.'[27] Epigraphic and literary records show that the institution of time keeping with the water clock and announcing it by means of the gong was adopted through all the centuries by royal courts (of the Tughluqs, Mughals, Rajputs, and even petty jamindars up to the beginning of the twentieth century), and also at places of worship belonging to the Hindu, Jain and Muslim faiths.[28]

23. See 0.6 above.
24. The bowl is now in the Museum of the Clockmakers' Company in London.
25. For a photograph of the full ensemble of the bowl, vessel and gong, see Virendra Nath Sharma, 'Astronomical Instruments at Kota'.
26. I-Tsing, *A Record of the Buddhist Religion*, pp. 144-145.
27. Al-Bīrūnī, *Alberuni's India*, vol. 1, pp. 337-338.
28. Cf. S.R. Sarma, 'Astronomical Instruments in Mughal Miniatures'; idem, 'Indian Astronomical and Time-Measuring Instruments'; S.R. Sarma & Ishrat Alam, 'Announcing Time: The Unique Method at Hayatnagar, 1676'; reprinted in this Volume, pp. 136-42.

1.0 Common householders could not afford the permanent installation of a water clock in their houses, for it needed the constant attendance of at least two people, one to announce the time when the bowl sinks and another to lift the bowl and place it again upon the water.[29] But householders too required the water clock on special occasions like marriages, in order to know precisely the astrologically auspicious moment (*śubha-muhūrta* or *śubha-lagna*, or simply *muhūrta* or *lagna*). Usually the Purohita who performed the marriage brought the water clock with him and set it up ceremoniously in the client's house.[30]

1.1 The ritual connected with the setting up of the water clock and its invocation is described, albeit briefly, in an unpublished manuscript entitled *Ghaṭikāyantraghaṭanāvidhi*. This manuscript cites Nārada as the authority for this ritual. The extant version of the *Nāradasaṃhitā* (before 1365) does describe the ritual but the wording is somewhat different. Likewise Govinda Daivajña's *Pīyūṣadhārā* commentary (AD 1603) on his paternal uncle Rāma Daivajña's *Muhūrtacintāmaṇi* (AD 1600) and Kāśīnātha Upādhye's *Dharmasindhu* (AD 1790-91) describe the ritual with different wording. Unfortunately, the relevant passages in all these four sources are corrupt. But with the help of these sources, a hitherto unknown ritual connected with the water clock can be reconstructed. In the following pages, I shall first describe the ritual as gleaned from these passages. After that, I shall reproduce the text of the manuscript, and also three parallel passages from the other texts, suitably emended as far as possible, and provide translation in English.

29. Actually, one needs at least six persons, namely two for each watch of eight hours.
30. Cf. Edgar Thurston, *Ethnographic Notes in Southern India*, part II, p. 565: 'This form of time-measurer, made of half a cocoanut or copper, is still in use among native physicians, astrologers and others in Malabar. . . . At the present day it is used on the occasion of marriage among higher Hindu castes. The brahmin priest brings the cup, and places the bridegroom in charge of it. It is the duty of the latter to count the gadis (= *ghaṭīs*) until the time fixed for his entrance into the wedding-booth.'

1.2 The *Ghaṭikāyantraghaṭanāvidhi* is a small paper manuscript of just three folios, now deposited in the Sarasvatī Bhavana of Sampūrṇānanda Saṃskṛta Viśvavidyālaya at Varanasi. The title *Ghaṭikāyantraghaṭanāvidhi* is mentioned at the beginning of the work. There are in all thirteen verses. These are followed by the expression '*atha maṅgalāṣṭakam*'. But before providing these eight auspicious verses, the manuscript breaks off abruptly. There is no colophon, nor any other indications to identify the author, scribe or the date. However, because of its close relation to some texts which belong to the early years of the seventeenth century, it is possible that the text of the present manuscript (if not the manuscript itself) also belongs to the early seventeenth century.

1.3 The ritual consists of (i) setting up the basin (*kuṇḍa*) on a sacred ground; (ii) placing the bowl therein at sunrise or sunset; (iii) the *mantra* in praise of the water clock; (iv) prognostication (*phala*) according to the cardinal direction to which the bowl moves when placed on the surface of the water in the basin; (v) and prognostication according to the direction in which the bowl finally sinks; and (vi) the recitation of the so-called *pala-vṛttas*.

1.4 All the four texts begin with the measurements of the bowl and the definition of the perforation. Here all the texts are highly corrupt. The confusion we have already noticed in astronomical texts is multiplied here many times. Though the *Yajamāna* is not expected to fabricate the bowl of the water clock as part of the ritual, the dimensions of the bowl and its perforation became part of the ritual text. And this too was apparently recited in course of the ritual. In a story contained in the *Kathāratnākara* which Hemavijaya Gaṇin composed in AD 1600 in Ahmedabad, there is an account of a Brāhmaṇa setting up the water clock for telling the time of his daughter's marriage, which runs thus:

'The Brāhmaṇa, who is especially well-versed in the whole range of astral science, wore a forehead mark made of saffron and rice-grains—

'The round vessel is made of ten *palas* of copper. In the *ghaṭikā* [bowl] the height should be made of six *aṅgulas*. The diameter there should be made to the measure of twelve *aṅgulas*. The good cherish a water clock that holds sixty *palas* of water.

—dropped the bowl, made fully according to the aforementioned prescriptions, in a basin filled with clean water at the time of the setting of the divine Sun.[31]

The weight and the size of the bowl are not relevant to the story. Yet these are mentioned, but not the hole which plays a role in the story: it gets blocked—not by some dirt in the water of the basin, but by a rice grain that got itself detached from the Brāhmaṇa's forehead mark and fell into the bowl—and the auspicious moment of the marriage lapses.

1.5 While the ritual sometimes preserves archaic technical processes, frequent recitation in the ritual of passages that are not immediately relevant and therefore are not clearly understood can lead to their distortion as well.[32] Both these processes occur in the case of the *Ghaṭikā-yantra*. Even today, in temples at Mathura, the midnight hour of Kṛṣṇa's birth is measured by this water clock. Thus ritual preserves an archaic practice of time measuring. But the passages containing the measurements of the bowl get distorted by constant repetition, because the Brāhmaṇa priests do not produce the bowl themselves, and the measurements are of no interest to them. These distortions can be seen in ample measure in the four texts which will be presented below. The *Ghaṭikāyantra-ghaṭanāvidhi* and the *Nāradasaṃhitā* provide the dimensions of the bowl and its perforation, not once but twice, ostensibly as two alternative methods (*prakārāntara*) but actually the same method from two different sources, and both times the prescriptions are garbled. The *Dharmasindhu*, on the other hand, mixes up the

31. *Kathāratnākara*, pp. 539-540:
viśeṣato niḥśeṣajyotiḥśāstrakuśalo vinirmitakuṅkumataṇḍulatilakaḥ sa vipraḥ
daśatāmrapalāvartapātre vṛttīkṛte sati /
ghaṭikāyāṃ samutsedho vidhātavyaḥ ṣaḍaṅgulaḥ //
viṣkambhaṃ tatra kurvīta pramāṇād dvādaśāṅgulam /
ṣaṣṭ[y]ambhaḥpalapūreṇa ghaṭikā sadbhir iṣyate //
ityādiparipūrṇapramāṇopetaṃ ghaṭikāpātraṃ svacchanīrabhṛte kuṇḍe bhagavato bhānor astagamanasamaye mumoca /

32. For an engaging account of similar distortions in modern times, see Madhav M. Deshpande, 'Contextualizing the Eternal Language: Features of Priestly Sanskrit'.

prescriptions for the sinking bowl type and the outflow type. Nevertheless, these passages are of interest from the viewpoint of cultural history and the process of text transmission. In my translation of these passages, I shall try to explain how certain distortions may have taken place.

2.1 It is prescribed that the ground where the basin is to be set up should be sloping to the east and to the north and be smeared with cow dung. On this ground, some grains of rice are sprinkled and a jewel or a piece of gold is placed. The basin, of copper or clay, is placed upon the grains of rice. According to the *Nāradasaṃhitā*, the basin is wrapped in a pair of clothes. The *Pīyūṣadhārā* adds that it should be decorated with sandal paste and flowers. Then it is filled with clean water.

2.2 The measuring of time, i.e. the counting the *ghaṭīs*, starts either at sunrise or at sunset as the case may be. This moment is defined in our passages as 'when half of the Sun's orb has risen or set.' At this moment the bowl should be placed on the water in the basin. Before placing the bowl, Gaṇeśa and the Sun are worshipped, so also the personal deity and the teacher of the householder who is the bride's father.

2.3 While placing the bowl upon the water in the basin, the bowl is addressed with a sacred formula which is said to have been composed by Nārada. However, the text of the formula varies in all the four sources; the one cited in the *Pīyūṣadhārā* shows the greatest divergence. The original text may be that which occurs in the *Nāradasaṃhitā*; a variation of this can be seen in the *Ghaṭikāyantraghaṭanāvidhi* and an elaboration of the same in the *Pīyūṣadhārā*.

Nāradasaṃhitā:

> *mukhyaṃ tvam asi yantrāṇāṃ brahmaṇā nirmitā purā //*
> *bhavābhayāya dampatyoḥ kālasādhanakāraṇam /*

Dharmasindhu:

> *mukhyaṃ tvam asi yantrāṇaṃ brahmaṇā nirmite purā /*
> *bhava bhāvāya dampatyoḥ kālasādhanakāraṇam //*

Ghaṭikāyantraghaṭanāvidhi:

> *yantrāṇāṃ mukhyarūpāsi brahmaṇā nirmite ghaṭi /*
> *dampatyoḥ śubhakālāptihetave bhava śobhane //*

Pīyūṣadhārā:

yantrāṇāṃ mukhyayantraṃ tvam iti dhātrā purā kṛtam /
dampatyor āyuvṛddhyardhaṃ putrādidhanahetave /
jalayantraka me tasmād iṣṭasiddhiprado bhava //

2.4 When the bowl is placed on the water, it does not remain stationary where it is placed. It keeps turning until it settles at some place, generally towards the edge of the basin. The cardinal direction where it settles is said to be indicative of future portents. Likewise, the direction where it sinks is used for prognostication. Besides the *Ghaṭikāyantraghaṭanāvidhi*, only the *Dharmasindhu* contains this section. According to both the sources, the result is not beneficial if the bowl settles or sinks in any one of the following four directions, viz., south-east, south, south-west and north-west. Furthermore, if the bowl sinks in the west, it is also not beneficial; on the other hand, settling in this direction is said to be beneficial; it is even said that the girl becomes the favourite—obviously of the in-laws.

2.5 In the *Ghaṭikāyantraghaṭanāvidhi*, this prognostication is followed by three verses the contents of which do not have any apparent relation with one another or with the subject of the water clock. Fortunately, in the story from the *Kathāratnākara* which has been mentioned above, there is a reference to the recitation of '*pala-vṛttas*': 'The Brāhmaṇa placed the bowl of the water clock in a vessel containing clear water at the time of the setting of the divine Sun. Because he was busy reciting the *pala-vṛttas* such as '*mā kānte pakṣasyānte paryākāśe svāpsīḥ. . .*'[33] This is the first line of the first of the three verses in our manuscript. The story narrates that, after placing the bowl upon the water of the basin, the Brāhmaṇa recited this and similar *pala-vṛttas*. Therefore all the three verses in our manuscript must be *pala-vṛttas* that are recited after placing the perforated bowl upon the water in the basin. What then does the expression '*pala-vṛtta*' mean and why must these be recited ?

The first of these three verses occurs also in Bhāskara I's commentary on the *Āryabhaṭīya*.[34] There Bhāskara explains that the

33. *Kathāratnākara*, p. 540: *ghaṭikāpātraṃ svacchanīrabhṛte kuṇḍe bhagavato bhānor astagamanasamaye mumoca / mā kānte pakṣasyānte paryākaśe svāpsīḥ ityādi-pala-vṛtta-paṭhanato . . .*

34. *Āryabhaṭīya*, p. 175.

time taken to utter sixty long syllables (*guru-akṣaras*) is one *vināḍikā*, and then cites the first of our three verses which consists exactly of sixty long syllables. '*Pala*' being a synonym of *vināḍikā*, *pala-vṛtta* designates a verse consisting of sixty long syllables, the reciting of which takes one *pala* of time, i.e. twenty-four seconds.

Indeed all the three verses at the end of our manuscript are such '*pala-vṛttas*'. Why should they be recited? The *Ghaṭikā-yantra* can measure the period of one *ghaṭikā* of twenty-four minutes. For measuring smaller periods of time, it is likely that one recited the appropriate number of these *pala-vṛttas*.[35] It is possible that in the early seventeenth century when the *Kathāratnākara* was composed and also when the *Ghaṭikāyantraghaṭanāvidhi* may have been put together, the main function of these verses was forgotten and that they were just recited, without being employed for measuring the fractions of the *ghaṭikā*, just as one recited the verses about the measurements of the bowl and of its perforation. Nevertheless, the manuscript preserves three verses which were meant to measure time, no matter whether they were so used or not. The first verse is also the oldest; for it was quoted by Bhāskara I in his commentary which he completed in AD 629.[36] The second verse seeks the blessings of all the heavenly bodies for the couple and the third verse celebrates the ten incarnations of Viṣṇu. In Sanskrit prosody, such verses containing fifteen long syllables in each quarter are named variously as *Kāmakrīḍā, Līlākhela* and so on. The last two verses are benedictory in nature. It is quite appropriate to recite them in the context of a marriage ritual. But the first verse is of a different nature. It is not clear how it came to be connected with this ritual.

I now reproduce the four passages. Orthography and *sandhi* are silently corrected. Occasionally numerical expressions are followed by numeral symbols, e.g. *daśa* 10; such numbers are omitted. Wherever the wording has been emended, the original reading of

35. S.R. Sarma, 'Measuring Time with Long Syllables'.
36. Principal Vaman Shivaram Apte's *The Practical Sanskrit-English Dictionary*, revised and enlarged edition, Poona 1957, Appendix A: Sanskrit Prosody, p. 1, cites this verse to illustrate the metre *Līlākhela,* attributing it to the *Sarasvatīkaṇṭhābharaṇa*. But the verse does not occur in this text.

the manuscript (MS) or of the edition (Edn) is shown in the footnotes.

3.1.1 GHAṬIKĀYANTRAGHAṬANĀVIDHIḤ[37]

śrīgaṇeśāya namaḥ / atha ghaṭikāyantraghaṭanāvidhiḥ /

śuddhaṃ tāmravinirmitaṃ daśapalaiḥ[38] *pātraṃ ghaṭārdhākṛti*[39]
mūlād ūrdhvaṣaḍaṅgulaṃ samaghanaṃ vistārato dvādaśa /[40]
viddhaṃ svarṇaśalākayā tripacayā (?) māṣaikayā tadghaṭīṃ
mittvā[41] *vai palaṣaṣṭivārapatanāc*[42] *cet pūrita sā ghaṭī //1//*

atha prakārāntaram /

ardhodayaṃ vāstamayaṃ suvīkṣya
yantraṃ pradadyāj jalapūrṇapātre /
ṣaḍaṅgulotsedhasamaṃ suvṛttaṃ
kṛtaṃ mukhaṃ yad [d]virasāṅgulaṃ tat //2//
palais tu tāmrair daśabhir jalasya
pūrṇaṃ palaiḥ ṣaṣṭibhir ambuyantram /
satryaṃśamāṣatritayaiva vṛt[t]a-
śalākayā madhyamabhāgaviddham //3//
samudrasaṃkhyāṅguladīrghayopa-
deśān mayūraṃ naravānarādyam (!) /
gurvakṣaraiḥ khendumitair asus taiḥ
ṣaḍbhiḥ palaṃ tair ghaṭikā khaṣaḍbhiḥ[43] *//4//*

atha ghaṭīsthāpanabhūlakṣaṇam /

37. For kindly providing me with a xerox copy of this manuscript, I am grateful to Professor Vidya Niwas Misra, the then Kulapati of the Sampūrṇānanda Saṃskṛta Viśvavidyālaya, and to Dr B.N. Misra, the then Librarian of the Sarasvatī Bhavana.
38. MS: *rasa6palaiḥ.*
39. MS: *°kṛtiṃ.*
40. MS: *vistārataḥ saptakaṃ* 7.
41. MS: *bhittvā.*
42. MS: *°paṭhanāc.*
43. MS: *khaṣaṣṭhiḥ.*

prāgudagpravaṇe deśe gomayenopalepite /
mṛnāyaṃ (?) kuṇḍikāpātraṃ[44] *sthāpayed avraṇaṃ śubham //5//*
kuṅkumāktena sūtreṇa pariveṣṭya parasparam /
svacchena vāriṇā pūrṇaṃ taṇḍulānāṃ tathopari //6//
niścale salile sthāpyaṃ kuṇḍikāyāṃ jalopari /
sthāpayed ghaṭikāyantraṃ sūryabimbārdhadarśanāt /
gaṇeśārkau ca saṃpūjya guruṃ natveṣṭadevatām //7//

[atha] ghaṭīsthāpanamantraḥ /

yantrāṇāṃ mukhyarūpāsi brahmaṇā nirmite ghaṭi /
dampatyoḥ śubhakālāptihetave bhava śobhane //8//

atha prāgādimadhyāntaghaṭībhramaṇaphalam āha /

yadrūpaṃ bhramaṇaṃ karoti ghaṭikā prāgādimadhyaṃ kramāt
saubhāgyaṃ[45] *nidhanaṃ vadhūmṛtisamaṃ yuktā ca rogais tanuḥ*[46]*/*
kanyā vallabhatām upaiti[47] *gaṇikātulyā yadā(?)dimaṃ*
syāt sādhvī sutavittabandhusahitā madhyasthitāryapradā //9//
uttareśānapūrvāsu ghaṭī pūrṇā śubhapradā /
dikṣu śeṣāsu kanyāyāḥ magnā vaidhavyadā smṛtā //10//
mā kānte pakṣasyānte paryākāśe deśe svāpsīḥ
kāntaṃ vaktraṃ vṛttaṃ pūrṇaṃ candraṃ matvā rātrau cet /
kṣutkṣāmaḥ prāṭaṃś cetaś ceto rāhuḥ krūraḥ prādyāt[48]
tasmād dhvānte harmyasyānte śayyaikānte kartavyā //11//
mārtāṇḍas tārānāthaḥ kṣoṇīsūnuḥ sūnuś cendor
vāgīśo daityācāryaś chāyāputro rāhuḥ ketuḥ /
nakṣatrair aśvinyādyais tārāyuktaiś cābhiḥ sarve
kuryāsuḥ[49] *kalyāṇaṃ vo nityārogyaṃ lakṣmīm āyuḥ //12//*
lokakṣemayāsīn matsyaḥ kūrmaḥ kroḍaḥ puṃsiṃho[50]

44. MS: *kuṇḍikāyugmaṃ*, which is obviously wrong. Only one basin is needed. Probably the corruption stems from '*vastrayugma*' in *Nārada-saṃhitā* 92.
45. MS: *saubhāgyaṃ na.*
46. MS: *rogaiś canuḥ.*
47. MS: *upaiti ka.*
48. MS: *prādyāt krūraḥ.*
49. MS: *kuyāsuḥ.*
50. MS: *puṃsīghe.*

yo hrasvākāro rāmo[51] *rāmaḥ kṛṣṇo buddhaḥ*[52] *kalkī /*
evaṃ nānārūpaṃ nānākāraı̣ nānānāmānaṃ
yogidhyeyaṃ devaṃ devānāṃ[53] *vande 'haṃ govindam //13//*

atha maṅgalāṣṭakam /

3.1.2 Translation

Salutation to śrī Gaṇeśa.

Now the method of setting up the water clock.

A pure vessel, made of copper of ten *palas* in weight, of the shape of a hemisphere, measuring six *aṅgulas* from the bottom to the top, evenly dense, in width twelve *aṅgulas*; pierced by a golden needle, made of one *māṣa* increased by three (*tripacayā*?). After measuring with that vessel, if it sinks sixty times (or, if it is filled in sixty *palas* of time), then it is a [proper] water clock. //1//[54]

Now another method.[55]

After carefully observing the rise of the Sun's orb up to the half, or the setting of the same, the instrument (i.e. the bowl) should be placed in a basin filled with water. The bowl is so made that its height is equal to six *aṅgulas*, and the circular opening is twelve *aṅgulas* in diameter. //2//

The water clock (*ambu-yantra*) is made of ten *palas* of copper; it is filled by water in sixty *palas*.[56] It is pierced at the central portion (i.e. centre of the bottom) by a round needle made of three and one-third *māṣas* of gold. //3//

and 4 *aṅgulas* length;. . .[57] Ten long syllables (*gurvakṣaras*) make

51. MS: *romo*.
52. MS: *bauddhaḥ*.
53. MS: *devānāṃ devaṃ*.
54. The verse is highly corrupt; besides messing up the measurements, the last line telescopes two separate prescriptions, viz. *ṣaṣṭivārapatana* 'sinking sixty times in a nychthemeron' and *palaṣaṣṭyā pūraṇa* 'filling in a period of sixty *palas*'.
55. What follows it not exactly another method, but rather the same information from another source.
56. This can also mean 'it is filled by sixty *palas* of water'.
57. Here the verse is contaminated by some irrelevant material '*upadeśān mayūraṃ naravānarādyam*', apparently from *Sūryasiddhānta* 13.21b: *toyayantrakapālādyair mayūranaravānaraiḥ //*

one breath (*asu*); six of these make one *pala*; sixty of these make one *ghaṭikā*. //4//

Now the characteristics of the ground on which the water clock is to be set up.

On a ground, sloped to the east and north,[58] which has been smeared with cow-dung, a vessel called *kuṇḍa*, faultless (*avraṇa*) and auspicious, should be placed //5//

upon grains of rice and should be encircled with thread dyed in saffron; then it should be filled with clear water. //6//

The water clock (i.e. the bowl) should be placed on the placid water in the basin, when the Sun's orb is half visible, after worshipping Gaṇeśa and the Sun, and after bowing to the teacher and to the personal deity. //7//

The sacred formula (*mantra*) for placing the water clock:

O Water Clock, you have been created by Brahmā as the foremost among the time measuring instruments. O auspicious one, be the means for measuring the auspicious time for the wedding of the couple. //8//

Now he tells the fruit of the rotation of the bowl, starting from the east, etc., and ending in the middle.

According as the bowl rotates in cardinal directions from the east up to the middle of the basin, it causes respectively the good fortune of having the husband alive and devoted (*saubhāgya*), death, near death of the bride (*vadhūmṛtisama*), the body full of diseases, the girl becomes the favourite [of all], resembles a courtesan, (?) virtuous, endowed with sons, wealth and relatives. Staying in the middle, [the bowl] grants noble [sons]. //9//

If the bowl becomes full (*pūrṇā*) [and sinks] in the north, north-east, or in the east, it bestows auspiciousness; if it sinks (*magnā*) in the remaining directions, it is said to inflict widowhood on the girl. //10//

58. Sloping to the east and north is considered to be auspicious; cf. *Rāmāyaṇa* (vulgate edition), 2.99.24: *prāgudagpravaṇāṃ vediṃ viśālāṃ dīptapāvakām*. In its stead the Critical Edition chose a reading which makes no sense. 2.93.23:

prāgudakśravaṇāṃ vedīṃ viśālāṃ dīptapāvakām /
dadarśa bharatas tatra puṇyāṃ rāmaniveśane //

[Now the *pala*-verses]:

Do not, O pretty one, at the end of the bright fortnight, sleep at a place open to the sky. Should it turn night, the cruel Rāhu, starving with hunger and roaming hither and thither, may eat you up, taking your pretty round face for the full moon. Therefore, after darkness, make your bed at a secluded place inside the house. //11//

May the Sun, the Moon, Mars, Mercury, Jupiter, Venus, Saturn, Rāhu and Ketu, all these, together with the lunar mansions beginning with Aśvinī, and all these stars, produce auspiciousness, constant good health, prosperity, and longevity [for the couple]. // 12//

For the welfare of the world, there [manifested the incarnations of] the Fish, the Tortoise, the Boar, the Man-Lion, One who had a Short Stature, [*Paraśu*] Rāma, Rāma, Kṛṣṇa, Buddha and Kalkin. I bow to Govinda, the god of gods, who in this manner assumed diverse forms, diverse shapes and diverse names, and who is meditated upon by sages. //13//

Now the eight auspicious verses.

3.2.1 NĀRADASAṂHITĀ[59]

tal lagnaṃ jalayantreṇa dadyāj jyotiṣikottamaḥ /
ṣaḍaṅgulamitotsedhaṃ dvādaśāṅgulam āyatam //86//
kuryāt kapālavat tāmrapātraṃ tad daśabhiḥ palaiḥ /
pūrṇaṃ ṣaṣṭir jalapalaiḥ ṣaṣṭir majjati vāsare //87//
māṣatrayatryaṃśayutasvarṇavṛttaśalākayā [60] */*
caturaṅgulāyatayā[61] *tathā viddhaṃ parisphuṭam //88//*
kāryeṇābhyadhikaḥ (?) ṣaḍbhiḥ palais tāmrasya bhājanam /
dvādaśa mukhaviṣkambha utsedhaḥ ṣaḍbhir aṅgulaiḥ //89//
svarṇamāṣena vai kṛtvā caturaṅgulakātmakaḥ /
madhyabhāge tathā viddhā nāḍikā ghaṭikā smṛtā //90//
tāmrapātre jalaiḥ[62] *pūrṇe mṛtpātre vāthavā śubhe /*

59. *Nāradasaṃhitā*, 29.86-95, pp. 181-184.
60. Edn: *māṣamātratryaṃśayutaṃ svarṇa°*.
61. Edn: *caturbhir aṅgulair āpas.*
62. Edn: *jalaḥ.*

gandhapuṣpākṣataiḥ sārdham[63] *alaṃkṛtya prayatnataḥ //91//*
taṇḍulasthe svarṇayute vastrayugmena veṣṭite /
maṇḍalārdhodayaṃ vīkṣya raves tatra viniḥkṣipet //92//
mantreṇānena pūrvoktalakṣaṇaṃ yantram uttamam /
mukhyaṃ tvam asi yantrāṇāṃ brahmaṇā nirmitā purā //93//
bhavābhayāya dampatyoḥ kālasādhanakāraṇam /

3.2.2 Translation

The best of the astrologers should measure (*dadyāt*)[64] that auspicious moment by means of the water clock. With a height of six *aṅgulas*, with a width of twelve *aṅgulas*, //86//

let a copper bowl be made, like a hemisphere, with ten *palas* of weight. It is filled in the duration of sixty *palas* (or, with sixty *palas* of water), and sinks sixty times in a day and night.[65] //87//

It should be pierced with a circular gold needle of three and one-third *māṣas* in weight and four *aṅgulas* in length. Then it is accurate. //88//

A copper bowl should be made with more than six *palas* (sic!). The diameter of the opening is twelve and the height six *aṅgulas*. //89//

Having made with one (sic!) *māṣa* of gold a [needle that is] four *aṅgulas* [in length], [with that] when the bowl (*ghaṭikā*) is pierced thus in the middle, it is then known as the water clock (*nāḍikā*). //90//

In an auspicious copper basin, or in a clay basin, that has been filled with water, having decorated it with effort by means of sandal paste, flowers and coloured rice, //91//

the basin which is placed upon grains of rice, to which a gold piece is added and which is covered by a pair of clothes, one should place the bowl after having seen the rise of half of the Sun's orb. //92//

With this formula, one should deposit the best of the instruments, endowed with the aforementioned characteristics: 'You have been

63. Edn: *sārdhair*.
64. '*lagnaṃ dadyāt*' is an interesting expression.
65. See section 0.7 above.

created a long time ago by Brahmā as the foremost among the [time measuring] instruments. For the safety (*abhaya*) of the couple, you become the means of measuring the time [of their wedding].' //93//

3.3.1 Pīyūṣadhārā[66]

sa ceṣṭakālaḥ kathaṃ sādhanīya ity ata āha kaśyapaḥ /
evaṃ guṇagaṇān vīkṣya lagnaṃ niścitya yatnataḥ /
siddhāntoktena mārgeṇa lagnakālaṃ prasādhayet //1//
jalayantreṇa tallagnaṃ dadyāt tenārcito dvijaḥ /
mukhaṃ dvādaśabhir vṛttam[67] *aṅgulaiś ca ṣaḍ unnatam //2//*
ghaṭārdhavat tāmrapātraṃ kuryāt tad daśabhiḥ palaiḥ /
ṣaṣṭir majjaty[68] *ahorātre ghaṭikāpātram uttamam //3//*
māṣatrayatryaṃśayutasvarṇavṛttaśalākayā /
caturbhir aṅgulair āyatayā viddhaṃ sphuṭaṃ nyaset // 4//
raver ardhodayaṃ dṛṣṭvā vāpy ardhāstamayaṃ tathā /
pūrvoktalakṣaṇaṃ yantraṃ mantreṇānena niḥkṣipet //5//
niḥkṣipej jalapūrṇapātra ity arthaḥ // yad āha Nāradaḥ //
tāmrapātre jalaiḥ pūrṇe gandhapuṣpair alaṃkṛte /
taṇḍulasthe ratnayute śucibhūmāv ahaspateḥ[69] */*
maṇḍalārdhodayaṃ vīkṣya jalapātre viniḥkṣipet //6// iti//
mantram apy āha sa eva /
yantrāṇāṃ mukhyayantraṃ tvam iti dhātrā purā kṛtaṃ /
dampatyor āyuvṛddhyardhaṃ putrādidhanahetave /
jalayantraka me tasmād iṣṭasiddhiprado[70] *bhava //7// iti//*

3.3.2 Translation

How is that desired auspicious moment of time to be determined: in reply to this question, Kaśyapa spoke thus:

In this manner, after considering all the good points, and having chosen, with effort, the auspicious moment (*lagna*) according to the

66. *Pīyūṣadhārā*, p. 424. The verse numbers have been added.
67. Edn: *vṛttaṃ dvādaśabhir.*
68. Edn: *bhajed.*
69. Edn: *aharpateḥ.*
70. Edn: °*pradaṃ.*

method taught by the Siddhāntas, the time of that auspicious moment should be calculated (*prasādhayet*). //1//

Let the Brāhmaṇa, who had been honoured by him (i.e. the householder) measure that moment (*lagnaṃ dadyāt*) by means of a water clock.[71] Let a copper bowl be made with ten *palas'* weight, like a hemisphere, with the circular mouth measuring twelve *aṅgulas* in diameter and six *aṅgulas* in height. If it sinks sixty times in a day and night, it is the best water clock. //2-3//

The bowl that has been clearly pierced by a circular needle of gold, of three and one-third *māṣas'* weight and four *aṅgulas'* length, should be placed [on the water]. //4//

After having seen the rise of half of the Sun's orb, or the setting of the half likewise, the instrument having the aforementioned characteristics should be deposited, with this sacred formula. //5//

'Should be deposited' means 'in a basin filled with water'. Thus spoke Nārada:

'In a copper basin, which is filled with water, which is decorated with sandal paste and flowers, which is situated upon grains of rice on a pure ground, and which is endowed with jewels (*ratnayuta*), after noticing the rise of half of the Sun's orb, the bowl should be deposited.' //6//

He also taught the sacred formula:

'You have been created a long time ago by Brahmā as the foremost among the [time measuring] instruments. Therefore, for increasing the longevity of the couple and for conferring on them sons, wealth and the like, O water clock of mine, grant them the fulfilment of their desires.' //7//

3.4.1 DHARMASINDHUḤ[72]

atha lagnaghaṭīsthāpanam / daśapalamitatāmraghaṭitaṃ ṣaḍaṅgu-lonnataṃ dvādaśāṅgulavistṛtaṃ ghaṭīyantraṃ kuryād iti Sindhuḥ /

> *dvādaśārdhapalonmānaṃ caturbhiś caturaṅgulaiḥ /*
> *svarṇamāṣaiḥ kṛtacchidraṃ yāvat prasthajalaplutam //*

71. Note the distinction between the two expressions '*lagnaṃ prasādhayet*' and '*lagnaṃ dadyāt*'.
72. *Dharmasindhu*, pp. 510-511.

iti tu śrībhāgavate tṛtīyaskandha uktam / asyārthaḥ asītiguñjātmakaḥ karṣaḥ / asyaiva suvarṇasaṃjñā / karṣacatuṣṭayaṃ palam / tathā ca ṣaṭpalatāmraviracitaṃ pātraṃ viṃśatiguñjonmitasuvarṇanirmitacaturaṅguladīrghaśalākayā mūle kṛtacchidraṃ kuryāt / tena chidreṇa yāvat prasthaparimitaṃ jalaṃ praviśati tena ca prasthajalapūraṇena tat pātraṃ jale magnaṃ bhavati tat pātraṃ ghaṭīkālapramāṇam / tatra prasthamānaṃ tu ṣoḍaśapalātmakam /

palaṃ suvarṇāś catvāraḥ kuḍavaḥ prastham āḍhakam /
droṇaṃ ca khārikā ceti pūrvapūrvacaturguṇam//

ity ukteḥ / granthāntare caturmuṣṭiḥ kuḍavaś catvāraḥ kuḍavāḥ prastha iti / kecit ṣaṣṭisaṃkhyākaguruvarṇoccār[aṇ]e palasaṃjñakālaḥ ṣaṣṭipalakālo nāḍikety āhuḥ / evaṃ pramāṇīkṛtaṃ ghaṭīyantraṃ sūryamaṇḍalasyārdhodaye 'ste vā jalapūrṇe tāmrapātre mṛtpātre vā kṣipet / tatra mantraḥ /

mukhyaṃ tvam asi yantrāṇāṃ brahmaṇā nirmite purā /
bhava bhāvāya dampatyoḥ kālasādhanakāraṇam //

anena mantreṇa gaṇeśavaruṇapūjanapūrvakaṃ ghaṭīyantraṃ sthāpayet / evaṃ sthāpitā ghaṭī āgneyayāmyanairṛtavāyavyādidiggatā na śubhā / madhyasthitānyadiggatā ca śubhā / evam āgneyādipañcadikṣu pūrṇā na śubhā // iti ghaṭīvicāraḥ //

3.4.2 Translation

Now the setting up of the water clock [for measuring] the auspicious moment.

The *Sindhu* declares that the water clock should be made of ten *palas* of copper, six *aṅgulas* high and twelve *aṅgulas* wide.

'[A vessel made of] half of twelve *palas*' weight, in which a hole has been made [with a needle of] four *māṣas* of gold and four *aṅgulas* [in length], till it is filled by (?) one *prastha* of water.'

Thus it has has been said in the third *Skandha* of the sacred *Bhāgavata.*[73] Its meaning is [as follows]: Eighty *guñjas* make one *karṣa*. The same has the designation of *suvarṇa*. Four *karṣas* are one *pala*. Thus, a vessel should be made of six *palas* of copper; it

73. *Bhāgavata* 3.11.9.

should be pierced at the base by means of a needle made of twenty *guñjas'* weight of gold and four *aṅgulas* in length. Through this perforation, by the time a *prastha* measure of water enters, that bowl sinks in the water, because of the *prastha* measure of water that filled it. Then that vessel becomes the standard measure for the period of one *ghaṭī*. There the unit of one *prastha* contains sixteen *palas*.[74]

For it has been said: one *pala* is four *suvarṇas*; then *kuḍava, prastha, āḍhaka, droṇa* and *khārikā*, are respectively each four times the previous unit.

In another text, it has been said that four fistfuls are one *kuḍava*, four *kuḍavas* are one *prastha.*

Some others say that the time taken for uttering sixty long syllables is one *pala*, and that the duration of sixty *palas* is one *nāḍikā*. The water clock, thus calibrated, should be placed in a copper basin or clay basin, full of water, when half of the Sun's orb has risen or set. There this sacred formula is recited.

'You have been created long time ago by Brahmā as the foremost among the [time measuring] instruments. For the sake of the state of [their] becoming a married couple (*dampatyoḥ bhāvāya*), you be the means of measuring time.'

With this sacred formula, preceded by the worship of Gaṇeśa and Varuṇa, the bowl should be placed [on the water in the basin]. If the bowl thus placed moves to the south-east, south, south-west,

74. What is described in the *Bhāgavata* is the outflow type of water clock. Here the perforation is made by a gold needle of four *māṣas* (= twenty *guñjas*) in weight and four *aṅgulas* in length, cf. *Kauṭilīya Arthaśāstra* and *Jyotiṣkaraṇḍaka*, cited in n. 4 and n 5 above. The volume of the water discharged by this clock in a fixed period of time is one *prastha*. In this type of clock, water does not enter (*pra-viśati*) but flows out (*niḥsarati*). Similar confusion occurs elsewhere also. For example, while the *Jyotiṣkaraṇḍaka* describes the water clock of the outflow type, the commentator Malayagiri interprets the passage in the sense of the sinking bowl type. I shall discuss these two types elsewhere in greater detail. Finally, after a long and tiresome excursion into metrology, the *Dharmasindhu* comes to the conclusion that the water that enters into the vessel has the volume of one *prastha* which is equal to sixteen (*ṣoḍaśa*) *palas*. Recall Śrīpati's view that the bowl should hold sixty (*ṣaṣṭi*) *palas* of water!

or north-west of the basin, it is not auspicious. If it stays in the middle, or moves to other directions, it is auspicious. Likewise, if it fills [and sinks] in the five directions starting from the south-east, it is not auspicious. Thus the discussion of the water clock.

Bibliography

Al-Bīrūnī. *Alberuni's India*, tr. Edward C. Sachau. Reprint: New Delhi, 1983.

Āryabhaṭīya. *Āryabhaṭīya of Āryabhaṭa* with the Commentary of Bhāskara I and Someśvara, critically edited with Introduction and Appendices, by Kripa Shankar Shukla. Indian National Science Academy, New Delhi, 1976.

Bhāgavata. *The Bhāgavata (Śrimad Bhāgavata Mahāpurāṇa)*, critically edited by H.G. Shastri. B.J. Institute of Learning and Research, Ahmedabad, Volume 1 (Shandhas 1-3), 1996.

Brāhmasphuṭasiddhānta. Brahmagupta, *Brāhmasphuṭasiddhānta*, ed. Sudhakara Dvivedin with his own *ṭīkā*. Benares, 1902.

Deshpande, Madhav M. 'Contextualizing the Eternal Language: Features of Priestly Sanskrit' in: Jan E.M. Houben (ed), *Ideology and Status of Sanskrit: Contributions to the History of the Sanskrit Language*. Brill's Indological Library, Volume 13. E.J. Brill, Leiden, 1996, pp. 401-436.

Dharmasindhu. Kāśīnātha, *Dharmasindhu*, with the commentaries by Vaśiṣṭha Datta Miśra and Sudāma Miśra. Kashi Sanskrit Series No. 183. Vārāṇasī, 1968.

Ghaṭikāyantraghaṭanāvidhiḥ. MS no. 37074, Sarasvatī Bhavana Library, Sampūrṇānanda Saṃskṛta Viśvavidyālaya, Varanasi.

Indian Astronomy: A Source-Book (based primarily on Sanskrit Texts), compiled by B.V. Subbarayappa & K.V. Sarma. Nehru Centre, Bombay, 1985.

I-Tsing. *A Record of the Buddhist Religion as practised in India and the Malay Archipelago* (A.D. *671-695*), tr. J. Takakusu. Oxford 1896. Reprint: Delhi, 1966.

Jyotiṣkaraṇḍaka. *Vallabhīyācāryīyaṃ Śrījyotiṣkaraṇḍakaṃ Prakīrṇakaṃ*, with the *Vṛtti* by Malayagiri. Ratlam, 1928.

Kathāratnākara. Hemavijaya Gaṇī, *Kathāratnakara*, published by Hiralal Hansraj. Jamnagar, 1911.

Kauṭilīya Arthaśāstra, ed. & tr. R.P. Kangle, vol. 1 (Bombay 1960). Second edition: Bombay, 1969.

Mahāsiddhānta. *The Pūrvagaṇita of Āryabhaṭa's (II) Mahāsiddhānta*, ed. & tr. Sreeramula Rajeswara Sarma. Marburg, 1966.

Nāradasaṃhitā. *Nāradamunipraṇītā Nāradasaṃhitā, sānvaya vimalā ṭīkā sahitā*, Vyākhyākār: Rāmajanma Miśra. Vārāṇasī, 1984.

Needham, Joseph. *Science and Civilisation in China*, vol. 3: Mathematics and the Sciences of the Heavens and the Earth. Cambridge, 1959.

Nṛsiṃha. *Siddhāntaśiromaṇi* of Bhāskara with his auto-commentary *Vāsanābhāṣya* and the *Vārttika* of Nṛsiṃha Daivajña, ed. Murali Dhara Chaturvedi. Sampūrṇānanda Saṃskṛta Viśvavidyālaya, Varanasi, 1981.

Pāṭīgaṇita. The Patiganita of Sridharacharya, with an ancient Sanskrit commentary, ed. & tr. Kripa Shankar Shukla. Lucknow University, Lucknow, 1959.

Pīyūṣadhārā. Rāma Daivajña, *Muhūrtacintāmaṇi,* with the *Pīyūṣadhārā* commentary by Govindajyotirvid and Hindi translation by Kedaradatta Jod. Second enlarged edition, Varanasi, 1979.

Sarma, Sreeramula Rajeswara. 'Astronomical Instruments in Mughal Miniatures.' *Studien zur Indologie und Iranistik,* 16-17, 1992, pp. 235-276; reprinted in this Volume, pp. 76-121.

———. 'Indian Astronomical and Time Measuring Instruments: A Catalogue in Preparation.' *Indian Journal of History of Science*, 29, 1994, pp. 507-528; reprinted in this Volume, pp 19-46.

———. 'The Bowl that sinks and Tells Time.' *India Magazine, of her People and Culture,* 14.9, September 1994, pp. 31-36; reprinted in this Volume, pp. 125-35.

———. 'Measuring Time with long Syllables: Bhāskara I's Commentary on Āryabhaṭīya, Kālakriyāpāda 2.' *Indian Journal of History of Science*, 36, 2001, pp. 51-54; reprinted in this Volume, pp. 143-46.

Sarma, S.R. & Ishrat Alam. 'Announcing Time: The Unique Method at Hayatnagar, 1676.' *Proceedings of the Indian History Congress*, 52nd Session, New Delhi, 1992, pp. 426-431, reprinted in this Volume, pp. 136-42.

Sharma, Virendra Nath. 'Astronomical Instruments at Kota.' *Indian Journal of History of Science,* 35, 2000, pp. 233-244.

Siddhāntaśekhara. Śrīpati, *Siddhāntaśekhara,* ed. Babuaji Misra. Calcutta, 1932-1947.

Siddhāntaśiromaṇi. Bhāskara, *Siddhāntaśiromaṇi,* with his auto-commentary *Vāsanābhāṣya* and the *Marīci* commentary of Munīśvara, ed. Dattātreya Āpaṭe, *Golādhyāya,* Anandasrama Sanskrit Series, No. 122. Poona, 1943, 1952.

Śiṣyadhīvṛddhidatantra. Śiṣyadhīvṛddhida Tantra of Lalla with the commentary of Mallikārjuna Sūri, ed. & tr. Bina Chatterjee. Indian National Science Academy, New Delhi, 1981.

Shukla, Kripa Shankar. 'Āryabhaṭa I's Astronomy with Midnight Day-Reckoning.' *Gaṇita* 18.1, June 1967, pp. 83-105.

Thurston, Edgar. *Ethnographic Notes in South India.* Reprint: Delhi, 1975.

PART III

THE ASTROLABE

9

Sulṭān, Sūri and the Astrolabe

0.1 In its construction, the astrolabe combines science, technology and art as no other product of the medieval world does. Hailed as the "King of Astronomical Instruments" in India,[1] and as "The Mathematical Jewell" in England[2]—these are the outermost limits of its geographical spread—, the astrolabe occupied in the medieval science a position analogous to that of the personal computer today. "Like a modern electronic computer," says John North, "the astrolabe in the Middle Ages was a source of astonishment and amusement, of annoyance and incomprehension. Imprecise as the astrolabe may have been in practice, it was undoubtedly useful, above all in judging time."[3]

0.2 It is not known precisely when the astrolabe was invented or by whom. But the principle of stereographic projection,[4] on which its construction is based, is attributed to Hipparchus who lived about 150 BC. By the sixth century AD, almost all the main components of the astrolabe were fully developed.[5] The Islamic

First published in: *Indian Journal of History of Science*, 35 (2000) 129-47.

1. Thus the Jaina monk Mahendra Sūri in AD 1370, see 6.1 below.
2. So reads the title of John Blagrave's book on the astrolabe published from London in 1585.
3. J.D. North, "The Astrolabe," *Scientific American,* 230 (January 1974) 96-106, esp. 106; reprinted in: idem, *Stars, Minds and Fate: Essays in Ancient and Medieval Cosmology*, London 1989, pp. 211-220.
4. In a stereographic projection, every circle on the sphere is represented also as a circle on the plane of projection, and the angle between any pair of curves remains unchanged.
5. The earliest detailed description of the astrolabe, which is still available, emanates from John Philoponus, who wrote an extensive work in Greek on the construction and use of the astrolabe in about AD 530. For an English translation of this work, see R.T. Gunther, *The Astrolabes of the World,* Oxford 1932, vol. I, pp. 61-81.

World, however, deserves the credit for preserving this knowledge, elaborating upon it and then disseminating it westwards up to England and eastwards up to India.[6]

0.3 The astrolabe (Arabic *asṭurlāb*), or more correctly the planispheric astrolabe (*asṭurlāb saṭḥī* or *musaṭṭaḥ*)[7] contains a two-dimensional projection of the heavens. Its principal parts are the following. The rete or spider (*'ankabūt*) is a perforated or open-work plate containing the ecliptic and a star map with pointers indicating the positions of some bright stars. Beneath the rete are to be found a series of plates or discs or tympans (singular *ṣafīḥa,* plural *ṣafā'iḥ*) which are specific to a particular terrestrial latitude and display stereographic projections of the local horizon, equal altitude circles or almucantars (*al-muqanṭara*), azimuth circles (*as-sumūt*), hour curves, etc. When the rete is correctly set and made to rotate upon the plate of a particular latitude, the astrolabe simulates the motion of the heavens above the localities situated on that latitude. The rete as well as the series of the plates are nested in the hollow space of a heavy circular plate with an upraised rim, called mater (*umm*). On the back of the mater, a diopter with two sighting vanes (*libna*) is pivoted to the centre. This diopter, styled alidade from *al-'iḍāda*, is the observational part of the instrument, with which the heights or altitudes of the heavenly bodies are measured.

1.1 When was the astrolabe introduced into India? In his *Indica,*

6. The literature on the astrolabe is voluminous. But the best technical introduction to the instrument can be found in Willy Hartner, "The Principles and Use of the Astrolabe" and "Asṭurlāb" reprinted in his *Oriens-Occidens*, vol. I, Hildesheim 1985, pp. 287-318; and J.D. North, op. cit. For an important Indian contribution to the literature, see M.P. Khareghat, *Astrolabes* (M.P. Khareghat Memorial Volume II), ed. Dinshaw D. Kapadia, Bombay 1950. On the history of the astrolabe in India, see my "Astronomical Instruments in Mughal Miniatures," *Studien zur Indologie und Iranistik*, Hamburg, 16-17 (1992) 235-276, esp. 237-241; reprinted in this Volume, pp. 76-121, esp. 79-84; "Indian Astronomical and Time-Measuring Instruments: A Catalogue in Preparation," *Indian Journal of History of Science*, 29.4 (1994) 507-528, esp. 518-522; reprinted in this Volume, pp. 19-46, esp. 36-42; and "Yantrarāja: the Astrolabe in Sanskrit," *Indian Journal of History of Science*, 34.2 (1999) 145-158; reprinted in this Volume, pp. 240-56.
7. As distinct from two other late but insignificant variants, viz., the linear and the spherical astrolabes, cf. Hartner, op. cit., pp. 317-318.

al-Bīrūnī claims to have dictated a manual on the astrolabe in Sanskrit verse.[8] No manuscript of such a text has come down to us, nor is this claim taken seriously today.[9] Nevertheless, it is quite possible that he must have brought the astrolabe with him and explained its principles to his Hindu interlocutors at Multan. Al-Bīrūnī did write several tracts on the astrolabe in Arabic. The most celebrated is the *Exhaustive Study of the Possible Methods for the Construction of the Astrolabe*.[10] This text is still unpublished but a few parts have been studied and translated. E.S. Kennedy describes it thus:

> Amid the plethora of medieval treatises on the astrolabe, this is one of the few of real value. It describes in detail not only the construction of the standard astrolabe but also the special tools used in the process. Numerical tables are given for laying out the families of circles engraved on the plates fitting into the instrument. Descriptions are also given of the numerous unusual types of astrolabes that had already developed in Bīrūnī's time. As for the underlying theory, not only are the techniques and properties of the standard stereographic projection presented, but also those of non-stereographic and non-orthogonal mapping of the sphere upon the plane.[11]

1.2 Besides this, al-Bīrūnī devoted a number of other works to the discussion of various types of the astrolabe. For instance, the *Easy Method for the Correction of the Astrolabes and the Use of its Northern and Southern Compound Devices* describes a special type of composite astrolabe that can be used both for the northern and

8. *Alberuni's India*, tr. Edward Sachau, first Indian reprint: New Delhi 1964, vol. I, p. 137: "Most of their books are composed in *śloka*, in which I am now exercising myself, being occupied in composing for the Hindus a translation of the books of Euclid and of the Almagest, and dictating to them a treatise on the construction of the astrolabe, being simply guided herein by the desire of spreading science."
9. That there were serious limitations to al-Bīrūnī's knowledge of Sanskrit was demonstrated by David Pingree, "Al-Bīrūnī's Knowledge of Sanskrit Astronomical Texts" in: Peter J. Chelkowski (ed), *The Scholar and the Saint,* New York 1975, pp. 67-81.
10. For the full Arabic title and other bibliographical information, see Ahmad Saeed Khan, *A Bibliography of the Works of Abū'l-Raiḥān Al-Bīrūnī,* New Delhi 1982, no. 46, p. 21.
11. *Dictionary of Scientific Biography,* vol. II, New York 1970, pp. 152-153 (s. v. Al-Bīrūnī).

the southern celestial hemispheres.[12] Therefore, we will not be much wrong in assuming that the astrolabe was introduced into India in the eleventh century, most probably by al-Bīrūnī. There is no evidence, however, that Hindu or Jaina astronomers in northwestern India, or on the Gujarat coast, responded to this new device immediately.[13]

1.3 But after the establishment of the Delhi Sultanate, Muslim scholars migrated to India in great numbers from Central Asia and these must have brought astrolabes and used them in India for calendrical and astrological purposes. By the mid-fourteenth century, the instrument was sufficiently well known among the Muslim elite of northern India to be mentioned in a work of fiction. Ikhtisan, a minister of Sulṭān Muḥammad bin Tughluq Shāh (1324-1351), mentions the use of the astrolabe for measuring time and for determining auspicious moments in his romance *Basāṭīn al-Uns.*[14]

2.1 The use and the manufacture of the astrolabe received an impetus under Sulṭān Fīrūz Shāh Tughluq (1351-1388) who actively promoted exchanges and innovations in science and technology. He was deeply interested in medicine, natural sciences and technology. His canal building activities, the engineering skills employed in the transportation of the massive Aśokan columns,[15] his endowments for hospitals[16] and educational institutions have

12. Ahmad Saeed Khan, op. cit., no. 47, pp. 21-22.
13. Apparently some contemporaries of al-Bīrūnī believed that the astrolabe, as also the gnomon, the celestial globe and the armillary sphere, was invented in India. Al-Bīrūnī denounces this view in strong terms and explains the Greek origin of the instrument as well as its name. Cf. *The Exhaustive Treatise on Shadows by Abū al-Rayḥān Muḥammad b. Aḥmad al-Bīrūnī*, tr. E.S. Kennedy, Aleppo 1976, vol. I, pp. 111-112; Al-Bīrūnī, *In den Gärten der Wissenschaft: Ausgewählte Texte aus den Werken des muslimischen Universalgelehrten*, übersetzt und erläutert von Gotthard Strohmaier, Leipzig 1991, p. 102.
14. Iqtidar H. Siddiqui, "Basāṭīn al-Uns: A Source of Information on the Sultanate of Delhi Under the Early Tughluq Sultans," *Quarterly Journal of the Pakistan Historical Society,* 36.4 (1988) 293-302.
15. Cf. J.A. Page, *A Memoir on Kotla Firuz Shah, Delhi,* Memoirs of the Archaeological Survey of India, No. 52, Delhi 1937.
16. Cf. R.L. Verma, "Sulṭān Fīrūz Shāh Tughluq, the Medieval Indian Hakim" in: B.V. Subbarayappa (ed), *Scientific and Technological Exchanges between India and Soviet Central Asia (Medieval Period),* New Delhi 1985, pp. 140-147.

been extensively discussed by modern scholarship.[17] However, his interest in astronomy and astronomical instruments has not been properly studied, save for a few oft-repeated clichés. The reason for this neglect lies in the peculiar nature of the sources, and also the historians' general ignorance about the basic tools of astronomy and time-keeping.

2.2 Fīrūz was greatly interested in the astrolabe. Under his direction, astrolabes were manufactured probably for the first time in India. Contemporary chronicles like Shams-i-Sirāj 'Afīf's *Tārīkh-i Fīrūz-Shāhī* and the anonymous *Sīrat-i Fīrūz-Shāhī* make frequent references to the Sulṭān's preoccupation with this instrument. The latter text, in particular, contains precious technical information about the manufacture of astrolabes at his court. This work was completed in 772 AH / AD 1370, and survives in a unique manuscript which was copied in December 1598, and which is now preserved in the Khuda Bakhsh Oriental Public Library of Patna.

2.3 Unfortunately, however, the *Sīrat* has not been published so far. The late Professor Syed Hasan Askari undertook an English translation of this text. Regrettably, the section on the astrolabe, corresponding to ff. 151a-161a in the manuscript, has many imperfections. With the kind help of Janab S.A.K. Ghori Saheb, I have compared this translation with the Persian original and extracted some coherent data on the astrolabe manufacture, which I shall present in the following pages. The main purpose of this exercise is to demonstrate the singular importance of the *Sīrat* for the history of the astrolabe in India and, hopefully, thus to induce some competent Persianist to undertake an annotated translation of this section of the *Sīrat*.

3.1 This chronicle narrates that the Sulṭān had the following five astrolabes manufactured under his direct supervision:

(i) A silver astrolabe for the latitudes of the seven climates (*iqlīm*),
(ii) a north-south astrolabe in silver,
(iii) a north-south astrolabe in brass for the latitudes of the seven climates,

17. Cf. Mohammad Habib & Khaliq Ahmad Nizami (ed), *A Comprehensive History of India,* vol. V: *The Delhi Sultanat* (AD *1206-1526*), New Delhi 1970; reprint: 1982, pp. 585-604.

(iv) a northern astrolabe in gold and silver, and
(v) a very large and very splend d north-south astrolabe in brass, designated as *Asṭurlāb-i Fīrūz-Shāhī.*

3.2 Before going into the details regarding the last mentioned astrolabe, it will be necessary to say a few words on the types mentioned here. Classical geography divides the inhabited portion of the northern hemisphere into seven climes or climates in such a manner that the maximum duration of daylight at the middle of each climate is half an hour longer than in the previous one.[18] When the *Sīrat* says that the astrolabe nos. 1 and 3 are meant for the latitudes of the seven climates, it implies that the astrolabes contain a number of tympans (*ṣafā'iḥ*) for different latitudes falling in the seven climates, or at least one *ṣafīḥa* for each climate.

3.3 Again, three of the five astrolabes are said to be both northern and southern, while the fourth was only northern (and perhaps also the first one). In a northern astrolabe (*asṭurlāb shumālī*), the stereographic projections on the rete and on the tympans are so drawn that the north celestial pole becomes the centre of the astrolabe and the outermost periphery is formed by the Tropic of Capricorn. Therefore, on the rete of this astrolabe, the positions of only such stars can be shown that lie to the north of the Tropic of Capricorn. Those lying to the south and visible to an observer in the northern temperate zone of the earth can, however, be shown in the rete of the southern astrolabe (*asṭurlāb janūbī*), where the south celestial pole forms the centre and the Tropic of Cancer the outermost periphery.[19]

Quite early, Muslim astronomers thought of combining the features of both the types in one, so that it can display the stars belonging to both the northern and southern celestial hemispheres.[20] Such a combined or composite astrolabe is called *asṭurlāb shumālī wa janūbī.* As we have mentioned before, al-Bīrūnī composed a manual on this variety of astrolabe.

18. Cf. *'Āin-i Ākbari of Abul Fazl-i-'Allām,* vol. III, tr. H.S. Jarret, rev. Jadunath Sarkar, Calcutta 1948, pp. 50-125.
19. Cf. Yukio Ohashi, op. cit.
20. Cf. David A. King, "Astronomical Instrumentation in the Medieval near East" in: idem, *Islamic Astronomical Instruments,* London 1987, p. 5.

3.4 However, the practical utility of such a north-south astrolabe is not very great. First, unless the astrolabe is very large, it will be difficult to include on it many stars from both the hemispheres, but then the larger the astrolabe the greater the difficulty in handling it. Furthermore, double projections on the rete and the tympans will reduce the accuracy of the lines. Thus, both the southern astrolabe and the north-south astrolabe are theoretical curiosities without much practical relevance. This is reflected also in the extant astrolabes, almost all of which are northern. Not a single exclusively southern astrolabe has been noticed so far, while there may be one or two north-south astrolabes. Perhaps for this very reason of rarity, and perhaps also because its construction involves a certain amount of ingenuity, Fīrūz was fascinated by the north-south astrolabe.

3.5 The author of the *Sīrat* invests the manufacture of the north-south astrolabe with a legendary aura. The astronomers at the Sulṭān's court, says he, discovered once a north-south astrolabe manufactured by Alexander's scientists. Fīrūz studied this astrolabe carefully and decided to produce similar astrolabes himself. Be that as it may, the Sulṭān did not need the chance discovery of Alexander's north-south astrolabe. He must surely have had access to the much more recent treatises on the north-south astrolabes by al-Bīrūnī.

4.1 Having decided to produce a north-south astrolabe,[21] the chronicle goes on to say, the Sulṭān summoned wise philosophers (*hukāmā*), astronomers (*munajjamān*), arithmeticians (*muhāsabān*), geometricians (*muhandasān*) and craftsmen (*sannāa*) from all over Hindustan to assist him in his job.

4.2 This north-south astrolabe was to have such large dimensions that the conventional geometrical tools like the rulers (*mistar*), compasses (*parkarha*), set-squares ('*usbādah*) and the common engraving tools could not be employed in its manufacture. There-

21. The *Sīrat* consistently calls this north-south astrolabe "*tamm*", meaning "whole astrolabe". Literally, of course, a north-south astrolabe can be considered a "whole astrolabe" but in astrolabic literature "*tamm*" is a technical term used exclusively for astrolabes having 90 almucantar lines.

fore, new tools had to be improvised, as the following couplet in the *Sīrat* informs us:

> Every instrument that was available before him
> Was further refined in his own time.

4.3 With these new tools, a large north-south astrolabe was manufactured and was given the grand designation *Asṭurlāb-i Fīrūz-Shāhī*. The *Sīrat* provides a fairly long description of this instrument. When shorn of the usual hyperbole,[22] this description is valuable in the absence of the real astrolabe which is no more extant. The account occasionally reads as if it is a quotation from a manual. The description is also valuable because it was composed soon after the manufacture of the astrolabe. It will be shown presently that the astrolabe is dated 21 Sha'bān AH 771 while the book was completed a year later in AH 772.

4.4 The *Sīrat* begins with a conventional enumeration of the components of the astrolabe by saying that the *Asṭurlāb-i Fīrūz-Shāhī* consisted of twelve parts, viz., ring (*ḥalqa*), shackle (*'urwa*), cord (*mismār*), throne (*kursī*), mater (*umm*), outer rim (*ḥajra*), tympans (*ṣafā'iḥ*), rete (*'ankabūt*), alidade (*'iḍāda*), pin (*qutb*), sighting vanes (*libna*), horse-shaped wedge (*faras*).

4.5 Thereafter the *Sīrat* goes on to describe the back (*ẓahr*) of the astrolabe. It mentions the graphs for sines (*jayb*) and cosines (*qaus*), and the arcs of solar declination (*mayl-i aftāb*), which were presumably engraved in the two upper quadrants. Then three kinds of shadow squares are enumerated, with gnomons measuring respectively seven feet (*ẓill-i haft qadmī*); twelve digits (*ẓill-i asābi'a*); and six and half feet (*ẓill-i shashnim qadmī*). These are well known types, discussed among others by al-Bīrūnī,[23] and are actually tables of tangents and co-tangents for different angles. However, normally only the first two types of shadow squares are displayed in the lower half of the astrolabe.

22. In the *Catalogue of the Arabic and Persian Manuscripts in the Oriental Public Library at Bankipore,* vol. 7: Indian History, Patna 1921, second impression: Patna 1977, p. 33, Maulana Abdul Muqtadīr states that in this chronicle "A strong tendency to eulogy and exaggeration is shown throughout. The narrative is florid, overloaded with pious effusions, generally ending in a compliment to the king."
23. Khareghat, op. cit., pp. 7-10.

The back is also said to contain much astrological data in concentric semi-circles around the shadow squares. Here were shown the "limits" (*ḥudūd*), "faces" (*wujūh*), trigons (*muthallatha*), their regents, exaltations (*sharaf*), dejections (*hubūt*), etc.[24] The astrological data also included a fourfold division of the lunar mansions according to Indian astronomers, namely blind, one-eyed, both-eyed, and clear-sighted. This corresponds to the classification as *andha, mandākṣa, madhyākṣa* and *sulocana* in Sanskrit astrological texts.[25] Furthermore, the back of the astrolabe is said to contain "a table of the position of the constellations" from 12 *Sha'bān* 771 to 1 *Ramaḍān* 907.

4.6 On the front side (*wajh*), the astrolabe contained three tympans (*ṣafā'iḥ*). On each side of these three tympans and on the inner side of the mater, stereographic projections were drawn for seven different latitudes. Each of these latitudes pertained to a separate climate. The *Sīrat* furnishes the names of towns, their climates, latitudes and the duration of the maximum daylight, as shown in the table below.

Pl. No.	*Town*	*Climate*	*Latitude*	*Maximum Daylight Hours*
1a	Tilang	I	[18]	13;06
1b	Ajmer	II	26°	13;40
2a	Delhi	III	20;39°	13;05
2b	Mosul	IV	36°	14;30
3a	Bukhara	V	39°	14;48
3b	Kashghar	VI	44°	15;22
Umm	Bulghar	VII	49°	16;04

4.7 After mentioning the particulars for each plate, the *Sīrat* adds the names of other adjacent towns with their latitudes. Thus we have here a kind of geographical gazetteer of some sixty towns, among which only seventeen are Indian. Generally, the gazetteer is engraved on the inner side of the mater. But in the Fīrūz-Shāhī astrolabe the inner side of the mater contained a projection for the

24. For the explanation of these astrological terms, see Hartner, op. cit., pp. 304-306; Khareghat, op. cit., 11-15.
25. Cf. for example, Rāma Daivajña, *Muhūrtacintāmaṇi*, Bombay 1933, 2.22.

seventh climate. The gazetteer, on the other hand, is apparently distributed in seven groups on the seven projections pertaining to the different climates. That is to say, on the second side of the first tympan which is calibrated for the latitude of Ajmer, the following information is engraved: the name of the town (Ajmer), climate (II), latitude (26°), maximum duration of the daylight in hours (13;40), and also the names of nine other towns together with their latitudes which lie in the second climate. Likewise, on the first side of the second tympan, calibrated for Delhi in the third climate, the names and latitudes of ten other towns are mentioned. This seems to be a more meaningful method of recording the geographical gazetteer.[26] However, this part of the manuscript of the *Sīrat* is extremely faulty. The latitudes given for many towns are quite wide off the mark. This is so even in the case of Delhi, for which town the *Sīrat* assigns the latitude of 20;39° and the maximum daylight of 13;5 hours, while the correct figures as available in contemporary Sanskrit sources are 28;39° and 13:49,36 hours respectively.

4.8 Since this is a north-south astrolabe, the projections on the tympans were made both for the northern and southern celestial hemispheres. First of all, says the *Sīrat,* five concentric circles were described on the tympans to represent the five diurnal circles (*madārāt-i panigāna*) of the Equator, Capricorn, Aries, Libra and Cancer. The correct sequence should be Capricorn, Aries, Cancer, Libra, and Capricorn. In fact, the circles of Aries and Libra coincide with the Equator. Therefore, in effect, the five concentric circles represent successively the Tropic of Capricorn, Equator, Tropic of Cancer, Equator, Tropic of Capricorn.[27] Then the altitude circles (*muqanṭarāt*) were drawn both for the northern and southern celestial hemispheres.

4.9 Finally, the rete (*'ankabūt*). The *Sīrat* says that the positions of eighteen fixed stars were marked on it, according to their

26. There are at least two extant Indian astrolabes which follow this method: (i) Indo-Persian Astrolabe, dated AD 1616, in the collection of the Sampurnanand Sanskrit University, Varanasi and (ii) Gurumukhī astrolabe dated VS 1907/AD 1850 in a private collection in London.
27. Cf. the contemporary Sanskrit manual *Yantrarāja*, p. 60, where this sequence is given correctly. This text will be discussed in 5.1 ff. below.

coordinates on 12 *Sha'bān* 771 (= 10 March 1370)[28] and on 21 *Sha'bān* 839 (= 9 March 1436), "so that the astrolabe may be used for 136 years from the time of manufacture." We may recall that the table of constellations on the back of the astrolabe was valid from 12 *Sha'bān* 771 to 1 *Ramadān* 907 (= 9 March 1502). This is the duration of 136 lunar years.

4.10 Thus the *Sīrat* mentions three dates, separated from one another by 68 lunar (66 solar) years. The first is undoubtedly the date of the manufacture of this astrolabe and the 18 star positions marked on the rete correspond to their coordinates on 10 March 1370. But when the *Sīrat* adds that the positions on 21 *Sha'bān* 839 (= 9 March 1436) were also marked and that the instrument can be used for the 136 years starting from the time of manufacture, it seems to imply that the star positions were marked for three different dates: 10 March 1370; 9 March 1436; and 9 March 1502. Now it is most unusual to display three different positions for each of the 18 stars and thus to construct 54 star pointers. It is possible that the rete of the astrolabe displayed star positions only of the date of manufacture, viz. 10 March 1370, and the table engraved on the back of the astrolabe or the manual composed in Persian at this time explaining the construction and use of this astrolabe contained the co-ordinates of these 18 stars on the afore-mentioned three dates.[29] It appears that here also the anonymous author of the *Sīrat* is quoting indiscriminately from the inscriptions on the astrolabe as well as from the manual. This hunch is strengthened by his statement at the end of the section on astronomy and astrolabes, where he says:

> Here from every chapter something has been extracted and offered by way of samples. The full texts on this branch of knowledge, written under emperor's desire and also the astrolabes constructed under his direction are preserved in the royal library.

28. I have used CALH Calendar Conversion Programme, Historical Edition, version 1.2, November 1966, developed by Benno van Dalen for converting the *Hijrī* into Julian dates and Pancanga vers. 2.0 developed by M. Yano and M. Fushimi for converting the Indian *tithis.*
29. Due to the precession of the equinoxes, the longitudes of the stars increase by about 1° in 66 solar years. As will be shown below in 7.2. Mahendra Sūri accepts a rate of precession of 1° in 66 2/3 solar years

5.1 Sulṭān Fīrūz's real achievement, however, lay not in mastering the principles of the astrolabe, nor in getting it manufactured—though both are creditable in themselves—but in disseminating the science of the astrolabe in India among the Jainas and Hindus.[30] The Persian sources are silent on this aspect but fortunately there survives a Sanskrit text on the construction and use of the astrolabe, entitled *Yantrarāja,* which was produced in AD 1370 under the auspices of Fīrūz. This was the first ever manual on the astrolabe in Sanskrit and was composed by the Jaina monk Mahendra Sūri. His pupil Malayendu Sūri wrote a commentary on it.[31]

5.2 It is well known that Fīrūz got a number of Sanskrit texts on medicine, astronomy and astrology translated into Persian.[32] This could have been possible only if he had gathered at his court Jaina and Hindu scholars, besides Muslim scholars. The *Sīrat* reports about these translations,[33] but it does not mention that any Jainas or Hindus were associated with this activity. Only once does the *Tārīkh-i Fīrūz-Shāhī* state that the Sulṭān summoned Brahmins and Jainas to decipher the writing on the Aśokan pillar but none could do so.[34]

30. I mention the Jainas first because they had good relations with the Delhi Sulṭāns. They also acted as intermediaries in the intellectual exchanges between the Islamic learning on the one hand and Sanskrit scholarship on the other.
31. Mahendra Sūri's *Yantrarāja,* together with the commentary by Malayendu Sūri, ed. Kṛṣṇaśaṅkara Keśavarāma Raikva, Bombay 1936. On this text and others of this genre, see S.R. Sarma, "Yantrarāja: the Astrolabe in Sanskrit" (n. 6 above); see also Sadashiva L. Katre, "Sultān Firūz Shāh Tughluq: Royal Patron of a Contemporary Sanskrit work," *Journal of Indian History*, 45 (1967) 375-367.
32. One of the texts thus translated was Varāhamihira's *Bṛhatsaṃhitā*, and this translation is still extant. Cf. S. Farrukh Ali Jalali & S.M. Razaullah Ansari, "Persian Translation of Varāhamihira's Bṛhatsaṃhitā," *Studies in History of Medicine and Science*, 9 (1985) 161-170.
33. Cf. Sreeramula Rajeswara Sarma, "Translation of Scientific Texts under Sawai Jai Singh," *Sri Venkateswara University Oriental Journal*, 41 (1998) 67-87.
34. Cf. Sreeramula Rajeswara Sarma, "Palaeographic Notes," *Aligarh Journal of Oriental Studies,* 3.2 (1986) 125-140.

6.1 Be that as it may, Mahendra Sūri was so impressed by the versatile functions of the astrolabe that he called it *Yantrarāja,* the king of astronomical instruments. At the beginning of his work which is also styled *Yantrarāja*, he says that the Muslims have written many manuals on the astrolabe in their language. Having extracted their essence, just as one extracts nectar after churning the milky ocean, he is presenting this work in Sanskrit.[35] We do not know exactly what these Arabic/Persian sources were which Mahendra Sūri consulted, but they certainly must have included al-Bīrūnī's various Arabic writings on the astrolabe.

6.2 Malayendu informs us that his teacher Mahendra Sūri was the foremost astronomer at Fīrūz's court. He concludes his commentary on each of the five chapters with a common colophon stanza, which translates thus:

> Master Mahendra Sūri, the great Sūri, was the foremost (*prasṭha*) among all the astronomers at the court of the illustrious Fīrūz, the lord of the Muslims. In this commentary on the *Yantrarāja* composed by Malayendu Sūri, who is like the honey-bee at the lotus feet of the said master, the first chapter entitled *Gaṇitādhyāya* is concluded.[36]

6.3 Like the *Sīrat*, Malayendu's commentary also furnishes a geographical gazetteer of 77 towns together with their latitudes, of which some sixty are Indian.[37] Here a special mention is made of "Hissār-Fīrūzābād founded by the illustrious King Fīrūz [at the latitude] 29°48′." Sulṭān Fīrūz did indeed construct a fort at Hissar and called it Hisār-Fīrūzah around 1355, and from there on laid

35. *Yantrarāja*, 1.3.
36. Ibid, p. 54:

śrīperojaśakendrasarvagaṇakaprasṭho mahendraprabhur
jātaḥ sūrivaraḥ tadīyacaraṇāmbhojaikabhṛṅgadyutā /
sūriśrīmalayendunā viracite 'smin yantrarājāgame
vyākhyāne gaṇitābhidaḥ prathamako 'dhyāyaḥ samāptiṃ gataḥ //

The printed edition reads in *a*: *gaṇakaiḥ pṛṣṭo* which makes no sense. I follow the reading in Ms no. 37 of the Department of Sanskrit, Aligarh Muslim University.

37. *Yantrarāja,* pp. 18-19. These are tabulated and compared with Ulūgh Beg's lists by David Pingree, "History of Mathematical Astronomy in India," *Dictionary of Scientific Biography*, vol. XV, New York 1978, pp. 626-627.

down his famous canal system.[38] Delhi is mentioned under the twin names Dillī and Yoginīpura, and is assigned to the latitude 28°39′. Again, whenever a concrete example is needed to illustrate a rule, the latitude of Delhi is taken as the basis for such calculations.[39] All of this goes to show that Malayendu was at the court of Fīrūz, obviously in the company of his teacher Mahendra.

7.1 But the most significant fact is the following. In the *Yantrarāja*, Mahendra furnishes a catalogue of thirty-two astrolabe stars together with their longitudes and latitudes.[40] In the commentary, Malayendu adds that the epoch for this catalogue is "*Saṃvat* 1427 *Caitra sudi* 15 corresponding to the thirteenth lunar day of *Sha'bān* in the Arabic year 771."[41] Now

VS 1427 *Caitra sudi* 15 = 12 March 1370 Tuesday
AH 771 *Sha'bān* 13 = 11 March 1370 Monday.

Thus there is a difference of one day between the Vikrama and Hijrī dates. However, since the lunar day in Islamic calendar begins at sunset, 13 *Sha'bān* begins at the sunset on Monday and lasts up to the sunset of Tuesday. Therefore the two dates coincide during the daylight hours of Tuesday, 12 March 1370.

7.2 Furthermore, there are two separate tables in the commentary containing the co-ordinates for the 32 stars respectively for *Saṃvat* 1427 *Caitra sudi* 15 (= 12 March 1370) and *Saṃvat* 1494 *Śaka* 1359 *Caitra sudi* 15 (= 21 March 1437). In the second table the longitudes have an increment of 1° added for precession.[42] In this context, Mahendra Sūri declares that the rate of precession is $\frac{720}{800}$ minutes of the arc in one year, which works out to 1° in $66\frac{2}{3}$ years.[43] Now it becomes clear why the *Sīrat* speaks of different sets of star co-ordinates at an interval of roughly 66 years.

38. Cf. Mohammad Habib & Khaliq Ahmad Nizami (ed), *A Comprehensive History of India*, op. cit., vol. V, pp. 587-588.
39. See, for example, *Yantrarāja*, pp. 16-18.
40. Ibid, I. 22-39. Cf. David Pingree, op. cit., p. 628.
41. *Yantrarāja*, p. 35: *Saṃvat 1427 varṣe caitraśudi 15 . . . yathā taddine arabī terīṣaṃ 771 sabānacandra 13 abhūt.* The printed edition reads the lunar date wrongly as 213 while the Aligarh Ms reads it correctly as 13.
42. Ibid, pp. 36-43.
43. Ibid, 1. 40.

7.3 The *Sīrat* states, it may be recalled, that the star positions marked on the *Asṭurlab-i Fīrūz-Shāhī* pertain to 12 *Sha'bān* 771. I am unable, at present, to account for the discrepancy of one day between 12 *Sha'bān* of the *Sīrat* and the 13 *Sha'bān* of Malayendu Sūri. Nevertheless, this near identity between the epoch of Mahendra Sūri's star catalogue and that of the star positions engraved on the *Asṭurlāb-i Fīrūz-Shāhī* cannot have been accidental. If Fīrūz had merely encouraged Mahendra Sūri to write a book on the astrolabe, the latter could have chosen any date as the epoch. But if the Sulṭān commissioned his Muslim astronomers on the one hand and Mahendra Sūri (together with his pupil and assistant Malayendu Sūri) on the other to compose books respectively in Persian and Sanskrit on the astrolabe, and further had ordered that his artisans manufacture a grand astrolabe employing therein the star positions calculated jointly by the Muslim and Jaina astronomers, then and only then will it be possible to have an identical date for all the three activities.

7.4 Then again, while 12 or 13 *Sha'bān* has no special significance in the Islamic calendar, *Caitra sudi* 15 is the beginning of the new year in Vikrama calendar in the North Indian *purṇimānta* system. Therefore, there is a greater chance of the epoch date being chosen by Mahendra Sūri rather than by the Muslim astronomers.

7.5 Even if it was not, the coincidence in the dates shows that there was a systematic and close cooperation between Muslim and Jaina/Hindu astronomers at the Sulṭān's court. This would also suggest that manuals may have been composed in Persian and Sanskrit, and that astrolabes may have been manufactured accordingly with legends in both Arabic/Persian and Sanskrit. The Persian manuals and astrolabes are no more extant. What exists is only the garbled account of the *Sīrat.* But the best record that still survives of the Sulṭān's endeavours with the astrolabe is undoubtedly the Sanskrit text and commentary by the two Jaina Sūris, Mahendra and Malayendu.

8.1 We have seen that the *Sīrat* attaches great importance to the north-south composite astrolabe (*asṭurlāb shumālī wa janūbī*). Mahendra's *Yantrarāja* also dwells on this variant. It describes how to construct not only the northern astrolabes (*saumya-yantra*), but also the southern astrolabes (*yāmya-yantra*) as well as north-south composite astrolabes (*miśra-yantra*).[44]

8.2 In his commentary, Malayendu provides elaborate tables for constructing the north-south astrolabe. Here he supplies the eccentricities and radii, in northern and southern hemispheres, at six degree intervals of altitudes, for the following localities.[45]

S.No.	*Town*	*Latitude*	*Maximum Daylight in ghaṭīs*
1	Tilaṅga	18°	32;44
2	Tryambaka	21°	33;34
3	Aṇahillapattana	24°	33;48
4	Tīrabhukta	27°	34;10,54
5	Dillī	28;39°	34;34
6	Nepāla	31°	35;06

8.3 We have mentioned that Fīrūz got a number of technical books on medicine, astronomy, astrology translated from Sanskrit into Persian and that this activity must have involved a close cooperation between Muslim scholars and Jaina/Hindu Sanskritists. It is gratifying that this cooperation should bear fruit in the reverse direction as well, when Mahendra Sūri composed the manual on the astrolabe, after having extracted the essence from many Arabic/ Persian books on this subject. Neither the astrolabes manufactured at Fīrūz's command, nor the Persian manuals composed at his instance survive today. But his efforts at disseminating the science of the astrolabe bore two different kinds of fruits respectively among Muslims and Jainas/Hindus of India.

8.4 Fīrūz's first attempts to manufacture astrolabes was avidly emulated by the Muslims. Astrolabe manufacture reached its zenith under the Mughals, especially during the reigns of Shāh Jahān and Aurangzeb in the 17th century.[46] On the other hand, the

44. Ibid, Ch. 3.
45. Ibid, pp. 19-25.
46. See Sreeramula Rajeswara Sarma, "The Lahore Family of Astrolabists and their Ouvrage," *Studies in History of Medicine and Science*, 13.2 (1994), 205-224; reprinted in this Volume, pp. 199-222.

encouragement Fīrūz extended towards writing manuals on the astrolabe was taken up enthusiastically by Jainas/Hindus, who produced some fifteen works in the next four centuries.[47] Even so, Mahendra's *Yantrarāja* remained unsurpassed in popularity. Among the Indian *Jyotisis* it enjoyed the same kind of high reputation as Nasīr al-Dīn al-Ṭūsī's celebrated Persian manual on the astrolabe in twenty chapters, popularly known as the *Bist Bāb*, did among the Islamic astronomers.[48]

9.1 While the section on the astrolabes in the *Sīrat-i Fīrūz-Shāhī* is thus extremely valuable for the history of the astrolabe in India, only one sentence from this entire section gained certain amount of currency, generating a myth about Fīrūz's interest in the astrolabe. It is necessary to set this myth right. At the very beginning of the description of the *Asṭurlāb-i Fīrūz Shāhī*, the *Sīrat* states that this astrolabe "was fixed on the top of the highest minaret in Fīrūzābād." While describing the manuscript in his catalogue, Maulana Abdul Maqtadīr lays special emphasis on this sentence: "The author says that . . . an astrolabe invented by the emperor himself, was constructed by his order and placed on highest minārah of Fīrūzābād."[49] Since then, nearly every one who writes on Fīrūz Shāh Tugluq mentions this fact as if it is of great scientific significance. To quote one typical instance, where this statement is further,

47. Some of these texts were discussed in Sreeramula Rajeswara Sarma, "Astronomical Instruments in Mughal Miniatures," op. cit., pp. 238-241; see in this Volume, pp. 81-83.
48. This is evident from the large number of manuscripts of this text. Ninety-nine manuscripts are listed in David Pingree, *Census of the Exact Sciences in Sanskrit*, Series A, vol. 4, Philadelphia 1981, pp. 393-395; vol. 5, Philadelphia 1994, pp. 296-297. The one-hundredth manuscript, I am happy to report, is with the Aligarh Muslim University; cf. S.R. Sarma, "A Catalogue of Sanskrit Manuscripts. (A) Jyotiḥśāstra, preserved in the Department of Sanskrit, Aligarh Muslim University (Typescript), No. 37, dated *Saṃvat* 1910 *Jyeṣṭha* kṛ 1 (which date corresponds to 23 May 1853). Moreover, the entire fifth chapter of this work is engraved on a large astrolabe, now preserved in the Science Museum, London; cf. Sarma, "Yantrarāja: the Astrolabe in Sanskrit," p. 154, Fig. 3; see in this Volume, p. 255, Fig. 12.3.
49. *Catalogue of Arabic and Persian Manuscripts in the Oriental Public Library at Bankipore*, vol. 7: *Indian History*, op. cit., pp. 28-33.

embellished: "He got constructed a special type of astrolabe and fixed it on the highest minaret of Fīrūzābād, where it, along with some other instruments, served as an astronomical observation post."[50]

9.2 But from what has been said about the construction of the astrolabe, it must be obvious that it is not an automatic recording device, much less an "astronomical observation post," whatever this latter expression might mean. The astrolabe has to be held and manipulated by the astronomer. He holds it in his right hand and adjusts the alidade with the left hand in order to take the altitude of the sun in the daytime or that of a prominent star at night. Then he "feeds" the altitude so obtained into the astrolabe by rotating the star map to the appropriate position. Then the star map together with the latitude plate on which it rests simulates the heavens at the place of observation at that moment. Now the astronomer can take various kinds of readings from the front of the astrolabe. All of this, it must be reiterated, has to be done manually by the astronomer. Then again, as the *Sīrat* itself reports, this special astrolabe contains three latitude plates with projections for seven different towns (see 4.6 above). That is to say, this astrolabe was fabricated for use in different localities and not just in Delhi alone.

9.3 In these circumstances, setting the astrolabe upon, or suspending it from, a lofty minaret would by itself serve no purpose unless the astronomer also stands there at least to read it. Thus for a fourteenth century Sulṭān to fix a rare astrolabe on the top of the highest minaret of Fīrūzābād and subject it to the vagaries of the north Indian weather would be as pointless as for a twentieth century prime minister to install a personal computer atop the Qutb Minar. For both these instruments have to be read by the user and manipulated by hand. Moreover, like the personal computer today,

50. O.P. Jaggi, *History of Science and Technology in India,* vol. VII: *Science and Technology in Medieval India*, Delhi 1977, p. 44. The same sentence is reproduced, without any attribution, by Pervin T. Nasir, "Muslim Contributions in Astronomy: Astrolabe Making during the Mughal Period," *Papers presented at the International Conference of Science in Islamic Polity: Islamic Scientific Thought and Muslim Achievements in Science*, Islamabad 1983, vol. 2, pp. 55-61, esp. p. 58.

the astrolabe was an expensive and precious instrument of the Middle Ages, and it was handled with care.[51]

9.4 Therefore, we must seek a rational explanation for this dubious sentence and not take it literally and then, in order to substantiate it, hypothesize that the astrolabe invented by Fīrūz is of a special kind that took automatic recordings. There has never been such a variant in the history of the astrolabe nor does the *Sīrat* confirm such an interpretation. This chronicle merely states that it was just a plane astrolabe on which projections for the northern and southern celestial hemisphere were drawn. Such models were known to al-Bīrūnī and others.

10.1 A likely solution to this problem is to be found in another contemporary chronicle of Fīrūz's reign, namely, the *Tārīkh-i Fīrūz Shāhī* by Shams-i-Sirāj 'Afīf. This chronicle was written some eighteen years after the completion of the *Sīrat*. A comparison of the two chronicles suggests a possible rational answer to the conundrum. In Book IV, chapter 18 of the *Tārīkh*, 'Afīf speaks of the technical inventions and innovations introduced by Fīrūz.[52] In this connection, he mentions various types of standards or banners (*nishānah*). First there is the mention of a pair of very large standards. Each of these were made of one maund of iron and were tied to the elephants with thick ropes. These elephants marched on either side of Fīrūz when he went out for hunting and the standards were visible up to a distance of three *kos*. The *Tārīkh* then goes on to state that Fīrūz also caused a *nishānah* called *aṣturlāb* to be suspended near/from the golden minaret (*minārah-i zarrīn*). Thus what the Sulṭān caused to be suspended from a minaret cannot be the real astrolabe, the best and the most elaborate one he got made,

51. While describing the astrolabe manufacture at Isfahan, Sir John Chardin narrates how carefully the Persians treat the astrolabe: ". . . the Persians always keep astrolabes in cases or bags so that the air of Persia does not rust, dirty, or eat away the body, as it does in our Western countries; even among the common people each keeps his astrolabe like a jewel." Cited in: A.J. Turner, *The Time Museum, Catalogue of Collection,* vol. I, part 1: *Astrolabes, Astrolabe-related Instruments*, Rockford 1985, p. 25, n. 73.
52. *Tārīkh-i Fīrūz-Shāhī*, ed. Maulvi Wilayat Husain, Calcutta 1890, pp. 269-270.

but a *nishānah*, a banner on which the astrolabe was painted or depicted. Why did he do it? Obviously to proclaim his great love for the instrument. Where was the banner hung from? Not from any random minaret but from the *minārah-i zarrīn*, the golden minaret.

10.2 Before yet another myth is built up around this golden minaret, the *Tārīkh* fortunately explains what it was: it was the Aśokan pillar, which the Sulṭān got transported from Topra in 1356 and set up within his citadel at Fīrūzābād (i.e. the present-day Firoz Shah Kotla, not very far from the Indian National Science Academy) on the top of a three storeyed building erected specially for this purpose, next to his Friday mosque. 'Afīf repeatedly mentions that Fīrūz called this pillar his golden minaret.

10.3 This pillar may also have been the highest minar in Fīrūzābād then, but there is not much space on its top to set up an "observation post" nor for the Sulṭān and his astronomers to stand there and take readings with the astrolabe. All that the Sulṭān can do from the minaret is to hang a banner with an astrolabe depicted on it in order to proclaim that the astrolabe was his favourite scientific instrument, that he mastered the principles of its construction and use, and that he got several astrolabes manufactured in his kingdom. The Sulṭān was fascinated enough by the massive polished stone pillar that he erected a special building for setting up the pillar upon it. Antiquarian though he was, he had no way of knowing that the pillar was caused to be carved and inscribed by Rājā Priyadarśī Aśoka more than sixteen hundred years earlier, but he could not have chosen a more appropriate place for the symbolic act of displaying his astrolabe banner.

Acknowledgements

Mr Ishrat Alam has obtained for me a copy of the pages dealing with the astrolabe from the typescript of Professor Askari's unpublished English translation of the *Sīrat-i Fīrūz Shāhī*, for which I am thankful. I am highly obliged to Janab S.A.K. Ghori Saheb who very patiently helped me compare the original Persian of the *Sīrat-i Fīrūz Shāhi* with the English translation. While revising the paper, I had the benefit of discussing some Arabic and Persian technical terms with Dr S. Jabir Raza.

10

The Lahore Family of Astrolabists and their Ouvrage

1.1 In 1935, at the request of Dr Harald von Klüber of Berlin, Syed Sulaiman Nadvi put together a list of astrolabes and celestial globes manufactured by one Ḍiyā' al-Dīn of Lahore. Through this exercise, Nadvi brought to light the existence of a family of astrolabe makers at Lahore, who were active during the reigns of Humāyūn to Aurangzeb. Nadvi identified eight astrolabes and four celestial globes made by four different members of this family.[1]

1.2 Sixty years later, today we know of more than one hundred instruments signed by seven members of this family belonging to four successive generations, thanks to the efforts of Derek Price,[2] Alain Brieux and Francis Maddison,[3] Emilie Savage-Smith,[4] David King,[5] and myself.[6] Besides the one hundred and odd signed instruments, there are also several unsigned specimens which can be attributed to this family on stylistic grounds. These instruments are scattered in various parts of the world: India, Pakistan, Iraq, Kuwait, Egypt, Turkey, France, Germany, the Netherlands, Britain, Canada and the United States of America.[7] Thus in the entire history of scientific instrumentation in the Middle Ages there has been no other family comparable to this one, be it in the long continuous family tradition, be it in the immense quantum of work produced, or

First published in: *Studies in History of Medicine and Science*, 13.2 (1994) 205-24.

1. Nadvi (a).
2. CCA.
3. Brieux & Maddison.
4. ESS.
5. King (a), (b).
6. Sarma (b). See also Sarma (a) and (c).
7. A tentative list of these instruments is given in the Appendix.

in the artistic and technical excellence of production, or in the innovations in design.

1.3 In the course of preparing *A Descriptive Catalogue of Indian Astronomical and Time-Measuring Instruments,* I had occasion to study more than fifty astrolabes and celestial globes of this family which are deposited now in various public and private collections in India, USA, France, the Netherlands, and UK.[8] My project will take a few more years to complete and there is a fair chance of some more unpublished instruments of this family coming to light. Even so, it is time to make an interim assessment of the achievements of this family, and also to pose some interim questions. Before doing this, a few words are in order on the function and history of the astrolabe and celestial globe.

2.1 The astrolabe and the celestial globe are the two most important astronomical instruments invented in the Hellenistic antiquity. Together with Ptolemaic astronomy, these two instruments were adopted by the Islamic World, where they were preserved, elaborated upon and disseminated westwards to Europe and eastwards to India.

2.2 The celestial globe[9] (Arabic: *al-Kura*) consists of a spherical globe made usually of brass on which the celestial equator, ecliptic, tropics and other circles are plotted. Upon this grid are marked the positions of about 1020 fixed stars either according to the coordinates given by Ptolemy in his *Almagest,* or by Ulūgh Beg in his Tables,[10] or by any other subsequent astronomer.[11] Often, the

8. Cf. Sarma (f).
9. On the function and history of the celestial globe, see the excellent account in ESS.
10. Thus on a globe made in the year 1047/1626, now at the Khuda Bakhsh Oriental Public Library, Patna, Qā'im Muḥammad writes that he made use of Ulūgh Beg's coordinates after adding 3° for precision. Cf. ESS, p. 225.
11. For example, Muḥammad Faḍlullāh b. Muḥammad Murād b. Muḥammad Mūsā Āsṭurlābī Mutawaṭṭin Aurangabad plotted the star positions on a globe he made in 1223/1808 (now at the Salar Jung Museum, Hyderabad) according to the *Zij Jadīj Āṣafia* compiled by Ḥusain Khān Rizwī at Hyderabad. Ghulām Ḥusain Jaunpūrī, on the other hand, determined the star positions afresh through his own observations with a modern Hadley's Reflecting Sextant and marked these in 1816 on a celestial globe (now in a private collection at Aligarh). For a description of the last mentioned globe, cf. Ansari & Sarma.

figures of the twelve signs of the zodiac, and also the figures of another 36 constellations like the Great Bear, as conceived in the Hellenistic and Islamic traditions, are drawn with great ingenuity.

2.3 The celestial globe is mounted on a stand consisting of a horizontal ring and a vertical ring, these two acting as the local horizon and meridian respectively. When the axis of the globe is adjusted according to the local latitude, the hemisphere visible above the horizontal ring will resemble the starry sky above the viewer's latitude.[12] Thus the celestial globe makes an excellent teaching instrument.

3.1 The astrolabe[13] (Arabic: *asṭurlāb* or *aṣṭurlāb*), on the other hand, is a versatile instrument for observation and computation. Here the great circles and other circles, the star positions, etc., are drawn on a plane of two dimensions[14] by a method called stereographic projection which was known already to Hipparchus in the second century BC. In a strereographic projection, the circles are drawn as circles and the angle between any two circles remains the same on the two-dimensional plane. Since the projections drawn on the common astrolabe pertain to the northern celestial hemi-

12. This job is performed today by the Zeiss projector in the planetarium, and there are also some computer software programmes available which project the heavens on to the monitor for any given point of time at any latitude. Yet the celestial globe is a more convenient tool.
13. The literature on the astrolabe is voluminous. The best introduction is available in Hartner; North. See also Gunther.
14. Therefore, the astrolabe is more correctly called the planispheric astrolabe. About a thousand years later, attempts were made to draw astrolabic projections on a three-dimensional globe and also on a unidimensional staff, which were respectively styled spherical astrolabe and linear astrolabe. But these are mere curiosities and had no practical relevance. They cannot perform all the functions a plane astrolabe is expected to do. As against some two thousand and odd planispheric astrolabes that survive today, there is only one extant spherical astrolabe and not a single linear astrolabe. Even so some modern writers tend to speak of threefold classification of the astrolabe into (i) planispheric, (ii) spherical and (iii) linear, which is clearly a-historical. In this paper, the word astrolabe refers always to the plane astrolabe.

sphere, it can also be called the northern astrolabe (*asṭurlāb shumālī*).[15]

3.2 The astrolabe consists of two principal parts. First, the rete (Arabic: *'ankabūt*) which is an openwork circular disc, containing stereographic projections of the ecliptic and of the positions of some prominent fixed stars. Leaving the ecliptic and the star positions, the remaining part of the disc is cut out, so that the rete forms a sort of net through which the lines upon the tympan underneath can be read. The second part, called tympan, or plate or disc (Arabic: *ṣafīḥa*) represents the sky as seen from the observer's latitude. It contains stereographic projections of the equator, the tropics, and equal altitude curves (Arabic: *al-muqanṭara*) and azimuth lines. There are also curves to measure time, in seasonal hours, or in equal hours. Some Islamic astrolabes contain additional curves to indicate the prayer times. The rete, when rotated above the tympan, will simulate the motion of the heavens upon that particular latitude on any given day. Since the tympan is calibrated for a specific locality one needs to have several tympans to serve different terrestrial latitudes.

3.3 The rete and the set of tympans are accommodated in the hollow space of a thicker disc with a raised rim, which is named appropriately as the mater (Arabic: *umm*). On the inner side of the mater, there is usually a geographical gazetteer. The mater is surmounted by a crown-like triangular projection called *kursī*, to which are attached a shackle, a ring and a cord for suspending the astrolabe. Often the *kursī* and the rete are highly decorated.

3.4 The back of the mater (Arabic: *ẓahr*) is divided into four quadrants. Each of these contains various trigonometric scales and astrological tables. A dioptre called alidade (Arabic: *al-'iḍāda*) with two sighting vanes is pivoted to the centre of the mater at the back. This is the observational part of the astrolabe.

15. In theory of course one can also draw the projections of the southern celestial hemisphere and thus construct a southern astrolabe (*asṭurlāb janūbi*) or combine both kinds of projections on the same instrument and call it north-south astrolabe (*asṭurlāb shumālī wa janūbī*). But the last two versions are intellectual curiosities and have little practical relevance. In fact, among the existing astrolabes, there are not more than ten specimens of the last two types.

3.5 With the alidade, one measures the altitude of the sun in the daytime, or the altitude of some prominent fixed star at night. When the rete is adjusted according to these values, it simulates the starry heavens upon the observer's place. Then one can read off from the dial the time and also note the times of prayer. One can also directly read off the ascendant for that moment and the other three points on the ecliptic, without resorting to complicated calculations. The knowledge of the these four points on the ecliptic is essential for casting the horoscopes, or for determining the auspiciousness or otherwise of a given moment. More important, the astrolabe works as an analog computer, and can be used to solve a number of trigonometrical problems. The astrolabe can also be used in land surveying, for determining the heights or depths of objects and calculating the distances.

3.6 Finally, it was thought that when one held the astrolabe in one's hand, one was in fact holding [the secrets of] the universe. Therefore any person who owned or understood the astrolabe was highly esteemed throughout the medieval world.

4.1 The astrolabe was introduced into India probably in the first half of the eleventh century by al-Bīrūnī, who authored several tracts on this instrument. In the subsequent centuries, scholars migrating from Central Asia to the court of the Sulṭāns at Delhi must have brought with them their personal astrolabes together with their collection of books.

4.2 Manufacture of the astrolabe commenced at Delhi under the auspices of Fīrūz Shāh Tughluq in the second half of the fourteenth century.[16] The *Sīrat-i Fīrūz Shāhī*, an anonymous chronicle composed at his court has a long account on the astrolabes manufactured at Fīrūz's orders.[17] According to this chronicle, Fīrūz got constructed five astrolabes. The grandest of these is named *Asṭurlāb-i Fīrūz Shāhī*. It is said to have contained projections of both the northern and the southern celestial hemispheres (*asṭurlāb shumālī wa janūbī*). But none of these survive today.

16. Fīrūz also sponsored the composition of the first manual on the astrolabe in Sanskrit by the Jaina monk Mahendra Sūrī in 1370. The manual is entitled *Yantrarāja* (ed. K.K. Raikva, Bombay 1936).
17. This account is analysed in Sarma (g).

4.3 The celestial globe was relatively a latecomer to India. Although it was a standard instrument for teaching astronomy in the madrasas, it was not mentioned in Indian context until the time of Humāyūn. This monarch was adept at using both the astrolabe and the celestial globe. Two charming anecdotes speak about his interest in these instruments.[18] But no celestial globes manufactured at the time of Humāyūn have survived either.

5.1 With one exception,[19] all the earliest surviving astrolabes and celestial globes emanate from the Lahore family. Hence the importance of this family in the history of scientific instrumentation in India. The patriarch of the family is called Allāhdād.[20] Two astrolabes made by him survive today. The first is dated 1567 and is now at the Salar Jung Museum of Hyderabad. The second one is not dated. It is at the Museum of the History of Science, Oxford. On both these he signed his name as Ustād Allāhdād Asṭurlābī Lāhūrī, thus proclaiming himself to be a resident of Lahore. Allāhdād's son 'Īsá adds the soubriquet "Humāyūnī" to his father's name, implying that he was the astrolabe maker to Humāyūn.[21]

18. Cf. Sarma (c).
19. The earliest extant globe was made by one 'Alī Kashmīrī ibn Luqmān in 998/1589, but save this globe nothing is known about him; cf. ESS, pp. 223-224. According to Abū 'l-Fazl, Maulānā Maqṣūd Hirawī manufactured astrolabes and celestial globes for Humāyūn but these have not survived; cf. Nadvi, p. 602.
20. On the question whether the name should be read as Allāhdād or Ilāhdād, cf. Abbot, p. 146.
21. There has been much unnecessary and unproductive controversy about the meaning and significance of the epithet *Humāyūnī* which, in conjunction with *asṭurlābī* can only mean an "astrolabe maker to Humāyūn." Sulaiman Nadvi conjectured that Humāyūn may have invented a special kind of astrolabe called *Humāyūnī asṭurlāb*, and that the members of the Lahore family received the epithet *asṭurlābī Humāyūnī* for manufacturing such astrolabes. But all the hundred and odd surviving astrolabes belong to the conventional type. On the other hand, it is true that Allāhdād's only dated astrolabe was made in 1567, i.e. eleven years after Humāyūn's death. But this does not preclude Allāhdād's manufacturing other astrolabes in the life-time of Humāyūn. It is also perfectly legitimate that Allāhdād's descendants commemorate the royal patronage extended to their ancestor by calling themselves the descendants of the Āsṭurlābī Humāyūnī.

5.2 Allāhdād's descendants sign their names invariably along with those of their ancestors.[22] From this and from the dated specimens, we can trace the history of this family from 1567 to 1691. Allāhdād's son 'Īsá[23] is known through three astrolabes produced between the years 1600 and 1604. He had two sons, Qā'im Muḥammad and Muḥammad Muqīm.

5.3 The dates on the instruments made by Qā'im Muḥammad range from 1609 to 1637. Consequently his period of activity coincides with the reign of Jahāngīr. Qā'im was in fact a protégé of Nawāb Abūl Ḥasan, brother of Nūrjahān Begum.[24] His extant instruments include six astrolabes—two of these he made jointly with his brother Muqīm—and four celestial globes. He is the first member of the family whose celestial globes survive today. He is said to have perfected the art casting celestial globes in one piece through the *cire perdue* or lost wax process.[25]

5.4 Qā'im's younger brother Muḥammad Muqīm has the largest number of astrolabes to his credit. We know of some 37 astrolabes made during the half a century of creativity between the years 1609 and 1659. In addition, there are also some eight unsigned astrolabes which can be attributed to him on stylistic grounds. However, he does not seem to have produced many celestial globes. There is only one single globe which bears his signature.[26]

22. For example, the inscription on a celestial globe made by Ḍiyā' al-Dīn in 1064/1653 reads as follows: *'amal aqall al-'ibād Ḍiyā' al-Dīn Muḥammad ibn Qā'im Muḥammad ibn Mullā 'Īsá ibn Shaykh Allāhdād Āsṭurlābī Humāyūnī Lāhūrī san 1064 Hijrī*, "The work of the least of servants Ḍiyā' al-Dīn Muḥammad, son of Qā'im Muḥammad, son of Mullā 'Īsá, son of Shaykh Allāhdād Āsṭurlābī Humāyūnī Lāhūrī, dated the year 1064 Hijrī."
23. He is usually referred to as Mullā by his descendants, but occasionally also as Ḥāfiẓ.
24. Infra 9.2.
25. Infra 8.1.
26. Though we know of some 37 instruments signed by Muqīm and eight attributable to him, only 17 of these are dated. A decade-wise distribution of the dated instruments shows large gaps in between:

1609-1610	1 astrolabe
1611-1620	0 astrolabe
1621-1630	5 astrolabes

5.5 Qā'im had one son by name Ḍiyā' al-Dīn Muḥammad, who was the most prolific and versatile member of this family. He manufactured both astrolabes and celestial globes in great numbers; he also crafted some unusual varieties of astrolabes and celestial globes. Between 1645 and 1680 he produced some 32 astrolabes and 16 celestial globes.[27]

5.6 Muqīm had two sons, Ḥāmid and Jamāl al-Dīn. Ḥāmid's instruments are dated from 1628 to 1691. During this period he produced some 11 astrolabes and 2 globes.[28] Ḥāmid also copied a manuscript of Naṣīr al-Dīn Tūsī's well-known work on the astrolabe entitled *Bist Bāb* together with an anonymous commentary.[29] The

---

1631-1640	5 astrolabes	1 globe
1641-1650	4 astrolabes	
1651-1659	1 astrolabe	

Of course, the 20 undated and 8 attributable astrolabes could have been made during these gaps, but still many could have been lost.

27. The decade-wise distribution of Ḍiyā' al-Dīn's production, as given below, shows that his production is much low in the last decade from 1671 to 1680.

1645-1650	6 astrolabes	4 globes
1651-1660	11 astrolabes	6 globes
1661-1670	9 astrolabes	4 globes
1671-1680	2 astrolabes	2 globes

This cannot be attributed to his old age, for he made some of his finest and more complex pieces in this decade. This could only mean that many pieces produced in this decade are lost, or have not yet been identified in museums.

28. The initial date is problematic. Ḥāmid's first extant work, an astrolabe now at the Houghton Library, Harvard University, Cambridge, Mass., is dated 1038/1628. His next available instrument is a globe in the Whipple Museum, Cambridge, UK, and is dated 1065/1655. The gap between the two is 27 years. It is extremely surprising that no instrument manufactured during this long period came down to us.

29. Salar Jung Museum, Hyderabad, Ms No. 3877, copied by Ḥāmid Asṭurlābī b. Muḥammad Muqīm on 10 Sha'bān 1087/17 September 1678. Cf. *Catalogue of the Persian Manuscripts*, vol. IX, Salar Jung Museum and Library, Hyderabad 1988, under the no. 3877. Dr Rahmat Ali Khan, the Keeper of Manuscripts, has kindly drawn my attention to this manuscript.

other son Jamāl al-Dīn produced five astrolabes between the years 1666 and 1691.

5.7 While instrument making reached its highest pinnacle in Ḍiyā' al-Dīn Muḥammad's oeuvre, the decline also can be simultaneously witnessed in that of his two first cousins. The quantity of their production is meagre in comparison to Ḍiyā' al-Dīn's. In quality also there is a steep and inexplicable degeneration. With these three cousins, instrument making in the family comes to end in 1691. If their descendants continued making instruments, none of them seems to have survived.

6.1 After this introduction to the various members of this family and their chronology, we may now examine the main features of their work. Allāhdād's two extant astrolabes exhibit nearly all the features that are distinctive of the work not only of his family but also of the other astrolabists of India. Therefore, it is essential to define these features precisely and to trace the sources of inspiration. Modern writers classify all the Indian astrolabes with legends engraved on them in Arabic/Persian characters (as opposed to those with Sanskrit legends) as Indo-Persian astrolabes because, as they say, these were influenced in style by the Persian astrolabes of the Safavid period. Certain degree of similarity exists no doubt between the Persian and Indian astrolabes as regards the high *kursīs*, star pointers being joined by floral traceries and perhaps in having geographical gazetteers engraved in concentric circles.

6.2 But the differences are not inconsiderable either. In Indian astrolabes the *kursī* is generally pierced while it is solid and decorated on the surface in the Persian exemplars. In the Indian astrolabe, the rete contains more star pointers and the back has more scales. The arcs of the signs of the zodiac are equidistant in Indian astrolabes whereas they are stereographically projected in Persian astrolabes. More important, in Indian astrolabes, a graph of the meridian altitude of the sun is plotted upon the arcs of the zodiac signs, which is not the case in Persian astrolabes. Finally, in the Persian Safavid astrolabes, the entire surface is filled with fine ornamental engraving, the letters and numerals being engraved in high relief against a patterned background. In contrast, the engraving on the Indian astrolabes is plain and austere.[30] Therefore,

30. Cf. Turner, pp. 25-26.

we must seek elsewhere for possible prototypes of Allāhdād's astrolabes.

6.3 The two astrolabes signed by Allāhdād have high *kursīs*. But the *kursī* on the dated astrolabe at Hyderabad is solid with a geometrical pattern engraved on the front. The tracery on the rete exhibits an archaic pattern of tiger's claws. These two elements are not repeated in any of the astrolabes of the family. As against this, the other astrolabe at Oxford has a pierced *kursī* with a finely cut design and the rete has a floral pattern. These two features became the hallmark of the Lahore astrolabes.

On the back, the top left quadrant has a sine graph and the top right quadrant has equidistant arcs of the signs of the zodiac. However, neither of the astrolabes contains the graph for the meridian altitudes of the sun. This graph was introduced for the first time by his son 'Īsá. In the two lower quadrants, Allāhdād engraves shadow squares for 12 digits and 7 feet respectively, and, within these squares, incorporates elaborate astrological tables.

On the inner side of the mater, Allāhdād provides a geographical gazetteer, containing the names of cities, their latitudes, longitudes, *inḥirāf* and the duration of the longest day. The majority of the cities are from the Middle East, starting from Mecca at 21;40° and reaching up to Samarqand at 39;37°. But there are also a few Indian cities, and this number grows with each successive descendant. The astrolabe at Hyderabad contains a gazetteer of 96 cities. It promises but does not give the *inḥirāf* and the longest day; the cells meant for these values are left blank. In the astrolabe at Oxford, all these values are given for some 157 towns. Allāhdād's successors provide such gazetteers of towns, but give only the latitudes and longitudes.[31]

6.4 Since Allāhdād calls himself a Lāhūrī, one would expect that his astrolabes contain a tympan calibrated for the latitude of Lahore (roughly 32°), but none of the tympans in the two astrolabes relate to any specific Indian city. Therefore, neither astrolabe can be used with accuracy in India.[32]

31. Many of these coordinates are derived from Ulūgh Beg's tables, but those for the Indian towns may have been measured locally or derived from local traditions; cf. Kennedy; Gibbs & Saliba, appendix.
32. On all the tympans several curves are inlaid with silver, an unusual feature not repeated by any of his descendants.

6.5 To sum up, though these are the first surviving astrolabes made in India, the two astrolabes by Allāhdād do not display any feature specific to India (except the names of some Indian towns in the gazetteer). However, in the undated astrolabe at Oxford, beginnings of the stylistic peculiarities of Lahore school can be seen, which gradually developed in the works of his descendants. Since his descendants invariably recite their genealogy up to Allāhdād, it may be safely assumed that he migrated to Lahore and set up his workshop there at the behest of Humāyūn. He may have come from Samarqand and continued to make astrolabes specific to Samarqand but did not or could not adapt the design to Lahore or any other Indian city.

7.1 The three surviving astrolabes made by Allāhdād's son 'Īsá present a uniform appearance with a solid and multi-lobed *kursī*. The tracery on the rete is more delicate than in that of his father. At the back, the top right quadrant contains the arcs of the signs of the zodiac upon which the curves for unequal hours are projected. In the two astrolabes at Chicago, there is an additional graph for the meridian altitude of the sun at 31;50° i.e. the latitude Lahore. This is the first time this feature occurs in the astrolabe. Later on, his descendants began plotting two graphs for 27° and 32°, which are the latitudes of the two imperial capitals Agra (modern value 27;10°) and Lahore (31;37°).[33] 'Īsá reduces the astrological data to the minimum and displays just the correspondences between the twelve signs of the zodiac and the twenty-eight lunar mansions in two concentric semicircles. Thus 'Īsá lays the real foundation for what are typically Indo-Persian astrolabes.

7.2 These then are the main features of the astrolabes made in this family. We shall now briefly discuss the work of the other members of this family and highlight some unusual specimens. Muqīm, as we have noted, is a very prolific astrolabe maker. Though he made a large number of instruments, rarely does he repeat the design or even the size. This shows his virtuosity as instrument maker. This also indicates that he had a discerning clientele. (See Figs. 10.1 & 10.2).

33. On this graph, see especially Frank & Meyerhof, pp. 14-15; Gunther, vol. I, p. 184.

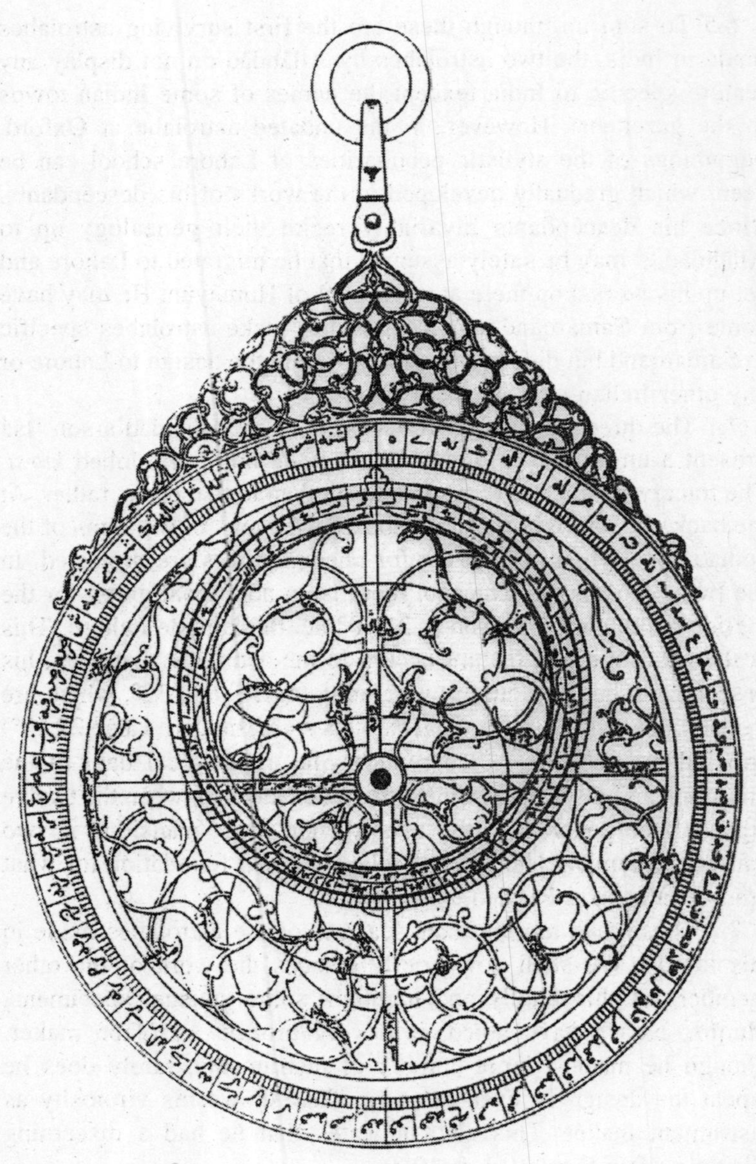

Fig. 10.1. Front of the astrolabe dated 1031/1621 by Muḥammad Muqīm.

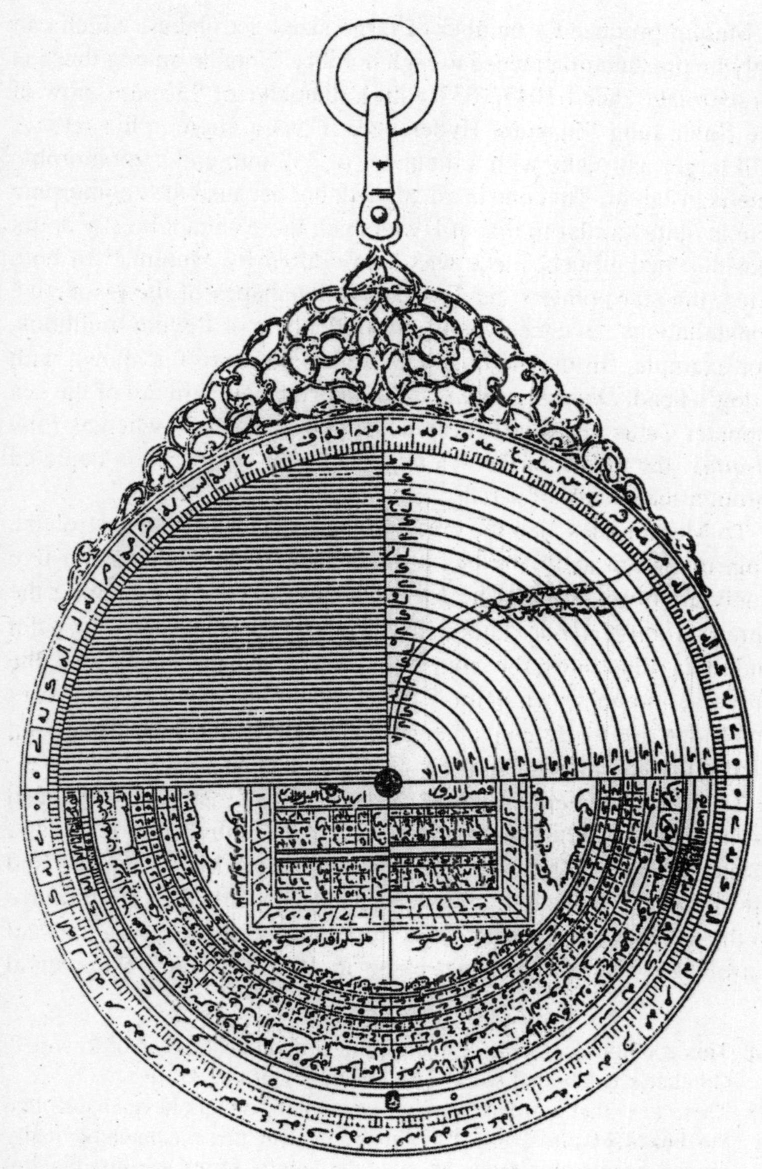

Fig. 10.2. Back of the astrolabe dated 1031/1621 by Muḥammad Muqīm.

Muqīm produced a number of large sized astrolabes, which can only be presentation pieces to high nobility. Notable among these is an astrolabe dated 1047/1637 with a diameter of 256 mm, now at the Salar Jung Museum, Hyderabad. It has a zoomorphic rete. A still larger astrolabe with a diameter of 352 mm and a zoomorphic rete is in Jaipur. This one is not signed, but because the zoomorphic rete is quite similar to that at Hyderabad, there cannot be any doubt that this magnificent piece was also crafted by Muqīm.[34] In both retes, the star pointers display the animal shapes of the respective constellations, as conceived in the Hellenistic or Beduin traditions. For example, Sirius, the dog star (α Canis Majoris) is shown with a dog's head. *Dhanab al-qīṭus al-janūbī* (the southern tail of the sea monster Cetus or β Ceti) is represented by a fish tail, whereas *Fam al-qīṭus* (the mouth of the sea monster Cetus or γ Ceti) is depicted through the mouth of a fish.[35]

To Muqīm goes also the credit of the world's smallest astrolabe. This miniature astrolabe has a diameter of just 43 mm, with five finely engraved plates. The *kursī* was cut *à jour* incorporating the phrase *Allāh-u Akbar*. This delicate piece fits nicely into the palm and thus illustrates the adage that the astrolabe represents the universe in one's own palm.[36] Neither the large ones nor this very small one are really convenient for actual use but must have been highly prized collector's items at the Mughal court.

7.3 Muqīm's nephew Ḍiyā' al-Dīn Muḥammad also produced several well crafted astrolabes with exquisite floral retes and thrones. Some of his retes display shapes of floral bouquets and are reminiscent of similar designs painted or inlaid in *pietre durre* in the Mughal monuments. Ḍiyā' al-Dīn also crafted some unusual astrolabes: a north-south astrolabe in 1085/1674 and a universal

34. This zoomorphic rete is reproduced in gold on the binding of Robert T. Gunther's celebrated work, *The Astrolabes of the World*.
35. There are other astrolabes where some star-pointers have shapes like the beaks of birds, etc. Cf. Gingerich. But these cannot be really termed zoomorphic retes, because the animal forms are just fanciful and do not represent the form in which the constellation is conceived. Therefore, these two astrolabes by Muqīm are the only true zoomorphic astrolabes.
36. Christie's Catalogue 24.9.1992, item no. 116, pp. 44-45.

Zarqālī astrolabe in 1091/1680. The universal astrolabe, designed originally by Ibn al-Zarqālluh in the eleventh century, does not require separate tympans for different localities. Ḍiyā' al-Dīn's Zarqālī universal astrolabe is a very large piece with a diameter of 555 mm.[37]

8.1 Aside from producing a large number of beautiful astrolabes, the Lahore family also made significant advances in the technique of manufacturing celestial globes. Until this time, the celestial globes were first made as two separate hemispheres and then joined together. But Qā'im Muḥammad developed the technique of casting them in one piece through the *cire perdue* or lost wax method, which is a highly complex process.[38] Qā'im's son Ḍiyā' al-Dīn excelled in the production of these seamless globes as well, and produced a large number of them.[39]

8.2 Though the basic design of the celestial globe remained the same from the Hellenistic times, Ḍiyā' al-Dīn did indeed introduce an innovation into one globe. Here he cut the surface of the globe *à jour*, as in the case of the rete of the astrolabe, leaving out the constellation forms and the great circles. At the star positions, he bored small holes. When lit from inside, the globe would present an illuminated celestial sphere, the stars shining through the perforations, and the constellations appearing in silhouette. Thus he applied the Mughal technique of the perforated brass lampshades to the ancient craft of globe making. This globe was commissioned by no less than the highest personage of the realm, i.e. emperor Aurangzeb himself.

9.1 With the exception of a few degenerate pieces by Ḥāmid and Jamāl al-Dīn, each one of the instruments manufactured by the various members of this family exhibits fine workmanship in metal,

37. Cf. Sarma (e).
38. ESS, pp. 90-95.
39. When Sulaiman Nadvi took his census in 1935, Aligarh boasted three instruments made by Ḍiyā' al-Dīn: two astrolabes dated 1064/1653 and 1074/1663 respectively and a globe of 1064/1663. The two astrolabes are not traceable today, but luckily the globe is still preserved in the Tibbiya College of Aligarh Muslim University, the only Indian university to own an instrument made in the Lahore family. For a detailed description of the globe, see Sarma (c).

meticulous engraving of the geometrical projections and a high degree of artistic excellence. Also, as A.J. Turner rightly puts it, "the multiplication of complexity and a delight in the unusual, seem to be typical of the astrolabe-makers of Lahore."[40]

9.2 Therefore these instruments appear to be have been sought after by nobility. In four cases their names are engraved on the instruments themselves. The earliest of these is a celestial globe which Qā'im Muḥammad made in the 18th regnal year of Jahāngīr (1032/1622) for Nawāb 'Itiqād Khān, who was a brother of Nūr Jahān Begum. The globe is now in Lancashire, England. Five years later, Qā'im designed an exquisite astrolabe for Nawāb Abūl Ḥasan, another brother of Nūr Jahān. Unfortunately, only the rete of this spectacular astrolabe survives at Patna. The star pointers on the rete are joined by a calligraphic design which states that the astrolabe was made in AH 1037 during the reign of Jahāngīr for Nawāb Jumdātul Mulk Khwājā Abūl Ḥasan. Perhaps this is the only astrolabe where calligraphy is fully incorporated into the design of the rete.

Qā'im's son Ḍiyā' al-Dīn designed, as we have stated above, an unusual celestial globe in 1090/1679 for Muḥī al-Dīn Muḥammad Aurangzīb Bahādur 'Ālamgīr. In the next year, i.e. in 1091/1680, he also designed a universal astrolabe for Nawāb Iftikār Khān, who was a Fauzdār of Jaunpur.

9.3 There are many other pieces which must have been produced for ostentatious display. Some of these magnificent pieces were acquired by the astronomer prince Sawai Jai Singh in the early eighteenth century and they formed part of his personal collection at Jaipur.[41]

9.4 The great demand for spectacular scientific instruments should have induced artisans outside this family also to produce astrolabes and celestial globes in equally large numbers. But surprisingly enough, very few instruments produced outside this family have survived. A prominent instrument maker outside this family is Muḥammad Ṣālih of Thatta who produced three fine astrolabes and four globes between the years 1665 and 1677.[42]

40. Turner, p. 83.
41. Cf. Sarma (d).
42. Cf. Sarma, Ansari & Kulkarni.

However, as against the nearly 130 instruments produced by the Lahore family in the late sixteenth and seventeenth centuries, those made by others do not even reach the number thirty. Surely the Lahore family was not granted a monopoly and others were prohibited from making instruments? The only possible explanation is that the ouvrage of the Lahore family overshadowed the work of others.

9.5 After Ḍiyā' al-Dīn, as we have mentioned, degeneration set in in the work of his first cousins and the production stopped altogether in 1691. Intriguingly enough, others also seem to have ceased producing astrolabes and celestial globes in the eighteenth century, for I have not come across a single instrument belonging to the eighteenth century, nor can I account for, at the present state of our knowledge, the sudden stoppage of production of Indo-Persian astrolabes and celestial globes at a time when Muḥammad Shāh at Delhi and Sawai Jai Singh at Jaipur were taking great interest in Islamic astronomy and astronomical instruments.

9.6 But this does not mean that the craft of the Lahore masters died at the end of the seventeenth century. In the nineteenth century, Lahore was once again the centre of production of astrolabes and globes. One Lālah Balhūmal, who proudly calls himself Lāhūrī, made astrolabes and globes of excellent workmanship.[43] Instrument-making spread also to other parts of India in this century. We hear of Zayn al-'Ābidīn making instruments in Delhi (none survive),[44] Ghulām Ḥusain Jaunpūrī at Tikari[45] and Muḥammad Faḍlullāh in Aurangabad.[46]

43. Cf. Anderson, p. 35, item no. 129a; ESS, pp. 52-54, 235-236, 244-245, 275-276, 304.
44. Sir Syed Ahmad, *Sīrat-i Farīdiya,* Urdu Ms. No. 10, University Collection, M.A. Library, A.M.U., Aligarh. *Editor's note*: Nawāb Zayn al-'Ābidīn Khān (d. 1856) was the younger son and pupil of Nawāb Farīduddīn Khān. He is said to be "an expert in making astronomical instruments and to have a profound knowledge of Zīj and astronomy (*hay'at*). Zayn al-'Ābidīn's own room was full of all sorts of instruments; for instance, he constructed a brass astrolabe and a brass sphere of quite large diameter", ibid, pp. 35, 43-44 (SMRA)—Ed.
45. Cf. Ansari & Sarma.
46. Cf. note 11 above.

9.7 The Lahore school of instrumentation, though itself dormant in the eighteenth century, inspired he production of astrolabes with Sanskrit legends for the use of Hindu Jyotiṣīs. In the eighteenth and nineteenth centuries a large number of Sanskrit astrolabes were produced in western India.[47]

47. On Sanskrit astrolabes, see Sarma (b), pp. 518-519.

BIBLIOGRAPHY

Abbot, Nabia: "Indian Astrolabe Makers," *Islamic Culture*, 11 (1937) 144-146.

Anderson, R.G.W.: *Science in India, A Festival of India Exhibition at the Science Museum, London, Catalogue*, London, 1982.

Ansari, S.M.R. & Sarma, S.R., "Ghulām Ḥusain Jaunpūrī's Encyclopaedia of Mathematics and Astronomy," *Studies in History of Medicine & Science*, 16.1 (1999-2000) 77-93.

Brieux, Alain & Maddison, Francis, avec la collaboration de Ludwik Kulus et Yusuf Ragheb: *Répertoire des Facteurs d'Astrolabes et leurs oeuvres: Islam, plus Byzance, Arménie, Géorgie et Inde Hindoue* (in preparation).

CCA—Gibbs, Sharon L., Henderson, Janice A. & Price, Derek de Sola: *A Computerized Checklist of Astrolabes,* Yale University, New Haven, 1973.

ESS—Savage-Smith, Emilie: *Islamicate Celestial Globes: Their History, Construction and Use,* Washington, D.C., 1985.

Frank, Joseph & Meyerhof, Max: *Ein Astrolab aus dem Indischem Moghulreiche*, Heidelberg, 1925.

Gibbs, Sharon & Saliba, George: *Planispheric Astrolabes from the National Museum of American History*, Washington, D.C., 1984.

Gingerich, Owen, "Zoomorphic Astrolabes and the Introduction of Arabic Star Names into Europe," *Annals of the New York Academy of Sciences*, 500 (1987) 89-104.

Gunther, Robert T.: *Astrolabes of the World*, Oxford, 1932.

Hartner, Willy: "The Principle and Use of the Astrolabe" in: idem, *Oriens Occidens*, vol. I, Hildesheim 1968, pp. 287-311.

Kennedy, E.S. & H.S.: *Geographical Coordinates of Localities from Islamic Sources*, Frankfurt am Main, 1987.

King, David A. (a): *Islamic Astronomical Instruments*, London, 1987.

King, David A. (b): "Medieval Astronomical Instruments: A Catalogue in Preparation," *Bulletin of the Scientific Instruments Society*, 31 (1991) 3-7.

Nadvi, Syed Sulaiman (a): "Some Indian Astrolabe-Makers," *Islamic Culture*, 9 (1935) 621-631.

Nadvi, Syed Sulaiman (b): "Indian Astrolabe-Makers," *Islamic Culture*, 11 (1937) 537-539.

North, John D.: "The Astrolabe," *The Scientific American*, 230 (January 1974) 96-106.

Raikva, K.K. (ed): *Yantrarāja of Mahendra Sūri*, Bombay, 1936.

Salar Jung Museum and Library: *Catalogue of the Persian Manuscripts*, vol. IX, Hyderabad, 1988.

Sarma, Sreeramula Rajeswara (a): "Astronomical Instruments in Mughal Miniatures," *Studien zur Indologie und Iranistik*, 16-17 (1992) 235-276; reprinted in this Volume, pp. 76-121.

Sarma, Sreeramula Rajeswara (b): "Indian Astronomical and Time-Measuring Instruments: A Catalogue in Preparation," *Indian J of History of Science*, 29.4 (1994) 507-528; reprinted in this Volume, pp. 19-46.

Sarma, Sreeramula Rajeswara (c): "From al-Kura to Bhagola: On the Dissemination of the Celestial Globe in India," *Studies in History of Science and Medicine* 13.1 (1994) 69-85; reprinted in this Volume, pp. 275-93.

Sarma, Sreeramula Rajeswara (d): "Portable Instruments at Jaipur Observatory," to be published in: S.M.R. Ansari & B.G. Sidharth (ed), *Proceedings of the International Symposium on Indian & Other Asiatic Astronomies,* Hyderabad- Jaipur 1991, Wiley Eastern, Delhi, 1995.

Sarma, Sreeramula Rajeswara (e): "Ṣafīḥa Zarqālīyya in India," *From Baghdad to Barcelona: Studies in the Islamic Exact Sciences in Honour of Prof. Juan Vernet*, Baralona, 1996, pp. 719-735; reprinted in this Volume, pp. 223-39.

Sarma, Sreeramula Rajeswara (f): "Indian Astronomical Instruments in the United Kingdom," paper read at the XIII Scientific Instrument Symposium, Leiden, September 1994.

Sarma, Sreeramula Rajeswara (g): "Sulṭān, Sūrī and the Astrolabe," *Indian Journal of History of Science*, 35.2 (2000) 129-147; reprinted in this Volume, pp. 179-98.

Sarma, S.R., Ansari, S.M.R. & Kulkarni, A.G.: "Two Mughal Celestial Globes," *Indian J of History of Science,* 28.1 (1993) 80-89; reprinted in this Volume, pp. 294-307.

Turner, A.J.: *The Time Museum Catalogue of Collections*, vol. I, part I: *Astrolabes, Astrolabe-related Instruments*, Rockford 1985.

APPENDIX

THE OUVRAGE OF THE LAHORE FAMILY
AN INTERIM CATALOGUE†

ASTROLABES

SHAYKH ALLĀHDĀD ASṬURLĀBĪ LĀHŪRĪ HUMĀYŪNĪ

1	* 975/1567	199—5	Hyderabad, Salar Jung M.	CCA 1120
2	* nd	256—6	Oxford, M. History of Science	CCA 1089

MULLĀ 'ĪSÁ IBN ALLĀHDĀD

3	*1009/1601	170—5	Chicago, Adler Planetarium	CCA 3823
4	*1013/1604	121—8	Chicago, Adler Planetarium	CCA 1076 = 3824
5	*nd	262—5	Ely, Cambs, UK, PC	CCA 68 = 3825

QĀ'IM MUḤAMMAD IBN 'ĪSÁ IBN ALLĀHDĀD

6	1034/1624	?—?	PLU. ex-Calcutta, Qāḍī 'Ubaydul Bārī	CCA 1128
7	*1037/1626	337—0	Patna, Khuda Bakhsh O.P. Library	
8	1041/1631	120—6	Baghdad, M. Arab Antiquities	CCA 3821
9	*1044/1634	193—7	Oxford, M. History of Science	CCA 71

QĀ'IM MUḤAMMAD & MUḤAMMAD MUQĪM

10	1018/1609	84—5	Hannover, Kestner M.	CCA 69
11	nd	125—5	Baghdad, M. Arab Antiquities	CCA 3820 = 3826

MUḤAMMAD MUQĪM

12	1031/1621	?—5	PLU. ex-Kazan, USSR	
13	*1032/1622	67.5—5	Paris, PC	

† The second column shows the date in AH/AD; the third column the diameter in mm and, in the case of the astrolabes, the number of tympans. In the fourth column is given the present or the last known location and in the fifth the serial number according to CCA or ESS. The following abbreviations are used. M = Museum; PC = Private Collection; PLU = Present Location Unknown. Those examined personally by me are marked with an asterisk.

14	1034/1624	87—5	Cengelköy, Kandilli Rasathane, Turkey	
15	1034/1624	114—5	Delhi, Red Fort M.	CCA 2700 = 3723
16	1034/1624	84—?	Samarqand, M. Culture and Art	
17	1047/1637	100—5	Brussels, PC	CCA 2601
18	1047/1637	260—?	Karachi, National M.	CCA 2704
19	1047/1637	203—5	Delhi, Red Fort M.	CCA 3721
20	*1047/1637	256—4	Hyderabad, Salar Jung M.	
21	1048/1638	102—?	Calcutta, National M.	CCA 3730
22	1050/1640	90—6	Brussels, PC, ex- Sottas	CCA 1119
23	*1051/1641	129—4	Oxford, M. History of Science	CCA 2531
24	*1051/1641	134—5	Greenwich, National Maritime M.	CCA 1054
25	*1053/1643	168—5	Oxford, M. History of Science	CCA 72
26	1053/1643	144—5	Washington, Smithsonian	CCA 86
27	*1070/1659	142—4	London, British M.	CCA 78
28	nd	203—5	Lahore, Museum	
29	*nd	130—5	Oxford, M. History of Science	CCA 1013
30	nd	140—5	PLU, Sotheby's	CCA 2609
31	nd	?—?	PLU, Sotheby's	
32	nd	43—?	PLU, Christie's	
33	*nd	91—4	Leiden, M. Boerhaave	CCA 1097
34	nd	90—6	Paris, PC	
35	*nd	105—5	Paris, Institut du Monde Arabe	CCA 3537
36	nd	204—5	PC, ex-Paris, Nicolas Landau	CCA 3529
37	nd	?—?	PLU, ex-Paris, Nicolas Landau	CCA 3827
38	nd	206—5	PLU, ex-Paris, Nicolas Landau	CCA 3828
39	nd	270—?	Paris, George Charliat	CCA 3504
40	nd	96—5	Cannes, PC	
41	nd	92—3	Fontanay-le-Comte, PC	
42	nd	147—5	Don Mills,Toronto, Science Centre	CCA 3837
43	*nd	134—0	New Haven, Yale University	CCA 3807
44	nd	90—5	Point Lookout, NY, Linton 166	
45	nd	92—5	Point Lookout, NY, Linton 223	
46	nd	?—?	Haifa, National Maritime M.	CCA 3565

ḌIYĀ' AL-DĪN MUḤAMMAD

47	1056/1646	193—6	PLU, ex-Paris, Nicolas Landau	CCA 2600
48	*1057/1647	121—4	Chicago, Adler Planetarium	CCA 1095
49	*1057/1647	117—4	Chicago, Adler Planetarium	CCA 2558
50	1059/1649	?—?	Lucknow, Nadwatul Ulema	CCA 1126
51	1060/1650	108—4	Brooklyn, Brooklyn M.	CCA 3555
52	1061/1650	210—5	PLU, Christie's	
53	*1062/1651	149—4	Cardiff, Welsh National M.	CCA 1107

54	*1062/1651	262—6	Hyderabad, Salar Jung M.	
55	*1064/1653	109—5	Oxford, M. History of Science	CCA 2533
56	1064/1653	?—?	PLU, ex. Habibganj	CCA 1118
57	*1067/1656	312—5	Jaipur, Observatory	CCA 2702
58	1068/1657	?—?	Cairo, M. Islamic Art	CCA 3829
59	*1068/1657	171—5	Paris, Jean Soustiel/Institut du Monde Arabe	
60	*1069/1658	177—5	Oxford, M. History of Science	CCA 77 = 1002
61	1070/1659	90—?	Washington, Smithsonian	CCA 87
62	*1071/1660	84—5	Chicago, Adler Planetarium	CCA 2554
63	1071/1660	114—0	PLU, ex-Rockford, Time M.	CCA 2607
64	*1072/1661	182—0	Paris, PC	CCA 3517
65	*1073/1662	106—4	Cambridge/Mass, PC	CCA 3809
66	*1073/1662	95—5	Chicago, Adler Planetarium	CCA 2551
67	*1074/1663	235—4	Patna, Khuda Bakhsh OP Library	CCA 1117
68	*1074/1663	130—5	Rampur, Raza Library	CCA 2511
69	*1074/1663	112—5	London, Victoria & Albert	CCA 1060
70	1074/1663	101—?	Mosul, al-Basha Mosque	
71	1074/1663	?—?	PLU, ex-Aligarh, Maulana Abū Bakr	CCA 1116
72	1077/1666	?—0	Karachi, National M.	
73	*1085/1674	165—1	Jaipur, Observatory	CCA 2703
74	*1091/1680	555—1	Jaipur, Observatory	CCA 80
75	nd	?—?	PLU, ex-Paris, Nicolas Landau	CCA 3524
76	nd	?—?	PLU, ex-Paris, Nicolas Landau	CCA 3525
77	nd	282—0	PLU, ex-Paris, Nicolas Landau	CCA 3651

ḌIYĀ’ AL-DĪN & ḤĀMID IBN MUQĪM

78	nd	?—?	PLU, ex-Paris, Brieux	CCA 3517

ḤĀMID IBN MUḤAMMAD MUQĪM

79	1038/1628	127—4	Cambridge, Mass, Harvard U.	CCA 3624
80	*1069/1658	112—5	Hyderabad, Salar Jung M.	
81	*1071/1661	112—0	Paris, Institut du Monde Arabe	CCA 3538
82	1084/1673	141—5	PLU, ex-Paris, Boisgirard 1977	
83	1086/1676	140—5	London, Aaron	CCA 3822
84	1087/1677	140—?	PLU, ex-Allahabad	
85	*1099/1688	89—3	London, Guildhall	
86	1102/1691	116—0	PLU, ex-Paris, Brieux	CCA 3517
87	1102/1691	126—6	PLU, ex-Paris, Jean Tetreau	
88	nd	110—?	PLU, ex-Renno Rizzi	
89	nd	?—?	Cambridge, Whipple M.	CCA 3563

JAMĀL AL-DĪN IBN MUḤAMMAD MUQĪM

90	1077/1666	248—?	Philadelphia, PC	
91	1077/1666	?—?	PLU, Christie's	
92	*1092/1681	168—?	London, Victoria & Albert	CCA 81
93	1094/1682	160—5	Kuwait, PC	
94	1103/1691	254—8	Istanbul, Türk ve Islâm Eserlieri Müzesi	

GLOBES

QĀ'IM MUḤAMMAD

95	1032/1622	188	Blackburn, Lanc, UK	ESS 11
96	1035/1625	?	Paris, PC	ESS 12
97	*J 22/1626	156	London, Victoria & Albert	ESS 13
98	*1046/1637	173	Patna, Khuda Bakhsh OP Library	ESS 14

MUḤAMMAD MUQĪM

99	1049/1639	?	Kuwait, PC / NY, Metropolitan	ESS 15

ḌIYĀ' AL-DĪN

100	*1055/1645	170	New York, Columbia U.	ESS 18
101	1057/1647	?	St. Petersburg, Asian M.	ESS 19
102	1058/1648	?	PLU, ex-Patna, PC	ESS 66
103	*1060/1650	113	London, Victoria & Albert	ESS 20
104	*1064/1653	122	Aligarh, Tibbiya College	ESS 21
105	*1067/1656	127	London, Victoria & Albert	ESS 22
106	*1068/1657	113	Cardiff, Welsh National M.	ESS 23
107	1068/1657	100	Cairo, M. Islamic Art	ESS 24
108	1070/1659	60	Cairo, M. Islamic Art	ESS 69
109	1071/1660	94	Berlin, Staatliche Museen	ESS 26
110	*1074/1663	160	Hyderabad, Salar Jung M.	
111	*1074/1663	142	Edinburgh, Royal Scottish M.	ESS 27
112	*1074/1663	175	Oxford, M. History of Science	ESS 28
113	1078/1667	?	Cairo, M. Islamic Art	
114	1087/1676	65	Delhi, Red Fort M.	ESS 71
115	*1090/1679	164	Rockford, Time M.	ESS 30

ḤĀMID IBN MUQĪM

116	*1065/1655	99.3	Cambridge, Whipple M.	ESS 68
117	*1094/1683	170	Hyderabad, Salar Jung M.	

ASTROLABE ATTRIBUTED TO ALLĀHDĀD

118 *nd	217—6	Oxford, M. History of Science	CCA 2530

ASTROLABES ATTRIBUTED TO MUQĪM

119 *nd	108—4	Oxford, M. History of Science	CCA 1014
120 nd	264—5	PLU, ex-Paris, Brieux	
121 nd	204—5	PLU, ex-Paris, Brieux	
122 nd	263—5	PLU, ex-Paris, Brieux	
123 *nd	240—5	Paris, PC	
124 nd	?—?	New York, Maxwell Rimler	
125 nd	217—4	Salem, Peabody M.	CCA 3555
126 *nd	352—7	Jaipur, Observatory	

GLOBE ATTRIBUTED TO QĀ'IM MUḤAMMAD

127 nd	217	Washington, Smithsonian	ESS 38

11

The Ṣafīḥa Zarqāliyya in India

To this volume in honour of Professor Juan Vernet, I am happy to make a small contribution that traces a connection between the scientific tradition of his country and mine. The mediator who brought about this connection is Ḍiyā' al-Dīn Muḥammad, the most prolific and versatile instrument maker of Mughal India. He is the great grandson of Allāhdād, the patriarch of the famous astrolabist family of Lahore.[1] Ḍiyā's father Qā'im Muḥammad developed the art of casting the celestial globes in one piece by the *cire perdue* method.[2] His uncle Muḥammad Muqīm is known through more than forty extant astrolabes[3] including the one that is considered the smallest in the world with a diameter of just four and half centimeters.[4] Ḍiyā' al-Dīn excelled in the production of both celestial globes and astrolabes. About forty five items crafted by him between the years 1637 and 1681 are extant today, scattered in various parts of the world.[5] His prolific work is also distinguished by superb craftsmanship and innovation in design.

For a celestial globe made for the emperor Aurangzeb in 1679, he invented an entirely new design. The surface of the globe was

First published in: Josep Casulleras and Julio Samsó (ed), *From Baghdad to Barcelona: Studies in the Islamic Exact Sciences in Honour of Juan Vernet*, Barcelona, 1996, pp. 718-35.

1. Cf. Sarma (iv); Savage-Smith, pp. 34-43.
2. On this technique, cf. Savage-Smith, pp. 91-95.
3. Cf. Gibbs *et al.*, *s.v.* M MUQIM B 6ISA; Brieux-Maddison, *s.v.* Muḥammad Muqīm b. 'Īsā; Sarma (iv), pp. 211-212, 215-219.
4. This tiny piece contains five finely engraved latitude plates as well, cf. Christie's Auction Catalogue, item 116, p. 44-45.
5. Cf. Gibbs *et al.*, *s.v.* DIYA AL-DIN M B KAIM M; Brieux-Maddison, *s. v.* Muḥammad b. Qā'im Muḥammad; Savage-Smith, *passim;* Sarma (iii); Sarma (iv).

cut *à jour*, as in the rete of the astrolabe, leaving out the outlines of the constellation figures, and various circles. Star positions were indicated by small perforations. When lit from inside, the globe would have presented an illuminated view of the celestial sphere.[6]

His astrolabes too exhibit considerable diversity. In size, they range from 84 to 555 mm in diameter. A typical example is the one at the Welsh National Museum, Cardiff. Made in 1062 AH/AD 1651, it has a diameter of 149 mm and contains four latitude plates.[7] Occasionally, he also made very large presentation pieces for the high nobility. A notable example of this type is deposited in the stores of the Jaipur Observatory.[8] It is a massive piece with a diameter of 312 mm and five latitude plates. On each plate the duration of maximum daylight is given both in hours and Indian *ghaṭīs*. Particularly noteworthy are the open work *kursī* and the jewel-like quality of the rete. This astrolabe is dated 1067/1656.

Besides this common variety of northern astrolabes, he also designed other types. For example, there is one astrolabe manufactured in 1674, on which both the northern and southern projections are displayed. This astrolabe, also from the stores of the Jaipur Observatory, is made of a single plate. The front is engraved as the latitude plate for 33° and the back for 35°. Moreover, the ecliptic circle is engraved on the front in its southern projection and on the back in the northern projection. On both sides, several stars are marked by dots enclosed in circles. A rete and a rule are pivoted to the front. The former consists of just the ecliptic circle, the meridian and a cross-bar. At the back, there is an alidade with a sighting tube attached to it. Unlike in his other astrolabes, the *kursī* here is plain and low.[9]

There exist four other unsigned north-south astrolabes of this type,[10] and all these can be attributed to Ḍiyā' al-Dīn. In particular, the astrolabe no. 809-1899 of the Victoria and Albert Museum,

6. Savage-Smith, pp. 42-43; 232-233; Fig. 17. The globe is now with the Time Museum, Rockford, IL.
7. Acc. No. 39.573/2.
8. Cf. Kaye, p. 16; Figs. 6, 8; Gunther, I, no. 75, p. 206, Figs. 104-106; Sarma (v).
9. Turner, p. 78, Fig. 58; Sarma (v).
10. Turner, pp. 74-83.

London, shows a close affinity with the one described just now. The London piece is also composed of a single plate, with an open work *kursi* at the top. The rete and the alidade are, however, missing. The front of the plate contains a double projection of almucantars, azimuths, hour lines and so on for the latitudes 32° and 36°. On these is superimposed the northern projection of the ecliptic circle. On the back there is a quadruple projection for the latitudes 22°, 24°, 25° and 29° and upon these is superimposed the ecliptic circle in its southern projection. Some fifty stars are marked on both sides with dots enclosed in small circles. A.J. Turner remarks that "the multiplication of complexity and a delight in the unusual seem to be typical of the astrolabe makers of Lahore".[11] This statement is particularly true of Ḍiyā' al-Dīn Muḥammad.

Sulṭān Fīrūz Shāh Tughluq of Delhi (reign 1351-1388), who promoted the manufacture of astrolabes and sponsored the composition, in 1370, of the first manual on the astrolabe in Sanskrit, entitled *Yantrarāja* by Mahendra Sūri,[12] is reported to have got made in the same year a gigantic north-south astrolabe and named it *Aṣṭurlāb-i Fīrūz Shāhī.*[13] This astrolabe is no more extant but Fīrūz's interest in this genre is reflected in nearly all the Sanskrit texts on the astrolabe.[14] This continued interest may also have resulted in the manufacture of some north-south astrolabes, but Ḍiyā' al-Dīn's are the only specimens that are extant.

Chronologically the last and technically the most remarkable of Ḍiyā' al-Dīn's productions is a universal Zarqālī astrolabe. The original prototype, as is well known, was invented by Ibn al-Zarqālluh (commonly known as Azarquiel) at Toledo in the eleventh century.[15] The astrolabe which he designed by making the vernal / autumnal equinox the centre and the solstitial colure the plane of

11. Turner, p. 83.
12. Cf. Raikva.
13. According to the anonymous *Sīrat-i Fīrūz Shāhī*, MS at the Khuda Bakhsh Oriental Public Library, Patna. For an analysis of the material in this manuscript concerning the astrolabes produced at the instance of Fīrūz, cf. Sarma (vii).
14. Thus, for example, Mahendra Sūri discusses the north-south astrolabe at Raikva, pp. 60-61.
15. Cf. Hartner, pp. 316-317; King; Puig (i); Turner, pp. 151-166.

projection obviates the necessity of having a separate plate or *ṣafīḥa* for each terrestrial latitude. Consequently this one consists only of a single plate, called *Ṣafīḥa Zarqāliyya* (or *Saphea Azarchelis*), on which are engraved the equatorial and ecliptic coordinates as well as some fixed stars. A movable ruler serves as the oblique horizon. This *ṣafīḥa* can be used at all latitudes and hence it is called universal.

The Zarqālī astrolabe made by Ḍiyā' al-Dīn is now part of a large collection of instruments deposited in the stores of the Jaipur Observatory erected by Sawai Jai Singh.[16] The inscription[17] on the astrolabe records that it was manufactured at Delhi for Nawāb Iftikār Khān in the twentythird regnal year of Emperor Aurangzeb in AH 1091, which corresponds to AD 1680-81.

It is a very large instrument, made out of a single sheet of brass, and measures 555 mm in diameter.[18] This large size naturally accomodates finer graduations and coordinates for shorter intervals. Thus the astrolabe is ideally suited for demonstration and computation rather than for actual observation. The *kursī* is cut from the same sheet as the astrolabe and bears simple ornamentation, with a pair of tulip-like flowers on both sides. There is a shackle, and a ring, both having diamond-shaped cross-section.

Originally the instrument must have been equipped with an alidade at the back for taking the altitudes, and a regula with a transversal cursor in the front for reading the coordinates of any point in either the equatorial system or in the ecliptic system. But

16. Garrett, p. 60, was the first to notice this instrument at the beginning of this century. He was not aware of its being a universal Zarqālī astrolabe but understood that it contained two sets of coordinates, and wrote: "This instrument was probably used as a star atlas, and also for the ready conversion of latitude and longitude into declination and right ascension and *vice versa*. Both sets of co-ordinates being engraved on the disc, one could be converted into the other by mere inspection, with an error of possibly a quarter of a degree". See also Kaye, pp. 27-30, 118, 124-125; Gunther, I, pp. 212-213; Savage-Smith, p. 43; Sarma (v).
17. The inscription is reproduced in Brieux-Maddison, *s.v.* Muḥammad b. Qā'im Muḥammad, no. 39.
18. The diameter is given wrongly by Kaye, p. 27, as two feet; Brieux-Maddison, *loc. cit.* and Savage-Smith, p. 43, as 610 mm.

these were missing in 1902 when the instrument was first noticed by Garrett.[19] Subsequently, some time before 1918, an alidade with a sighting tube in brass was mounted in the front, and cross-bars consisting of four arms at right angles, also in brass, were pivoted to the back.[20]

The Zarqālī projection is engraved on the front side of the instrument (see Fig. 11.1). Here the outer rim is graduated for every three degrees and numbered in *abjad* notation in clockwise direction, starting at the top. The next concentric ring is graduated for every degree and numbered in Arabic numerals, separately for each quadrant from 1 to 90, also in clockwise direction, starting at the top. Next there are divisions for every one-sixth of a degree. Inside these degree scales, two sets of coordinates are drawn: the equator with its circles of declination and right ascension, and the ecliptic with its corresponding circles of longitude and latitude. The circles of declination and those of latitude are drawn for every degree and numbered from the centre to the poles. The circles for right ascension and those for longitude are drawn for every three degrees. Dotted lines are used for the equatorial system and simple lines for the ecliptic system, so that these two can be clearly distinguished (see Fig. 11.2).

The equator is graduated for every three degrees and numbered, on the northern side from 1 to 180 running from the top to the bottom, and on the southern side from 181 to 360 in the reverse direction. The ecliptic is marked with the zodiacal signs from Capricorn to Gemini in the north and from Cancer to Sagittarius in the south, so that Aries and Pisces lie at the centre. Each sign is divided for every three degrees.

Several star positions are marked with dots enclosed in circles and labelled.[21] This celestial map is also made to serve

19. Thus in 1902, Garrett, p. 60, writes: "There is a round hole in the centre of the disk which probably indicates the former existence of some kind of sighting bar for taking altitudes, but if so, this has now been lost".
20. In 1918, Kaye, p. 27, speaks of the cross-bar but not of the alidade with the sighting tube. But the latter also must have existed then, for both are of the same workmanship.
21. Kaye, p. 118, gives the names and coordinates of only 14 fixed stars, but there are more. A full star list has yet to be made.

Fig. 11.1. The Zarqālī astrolabe by Ḍiyā' al-Dīn. Front view.

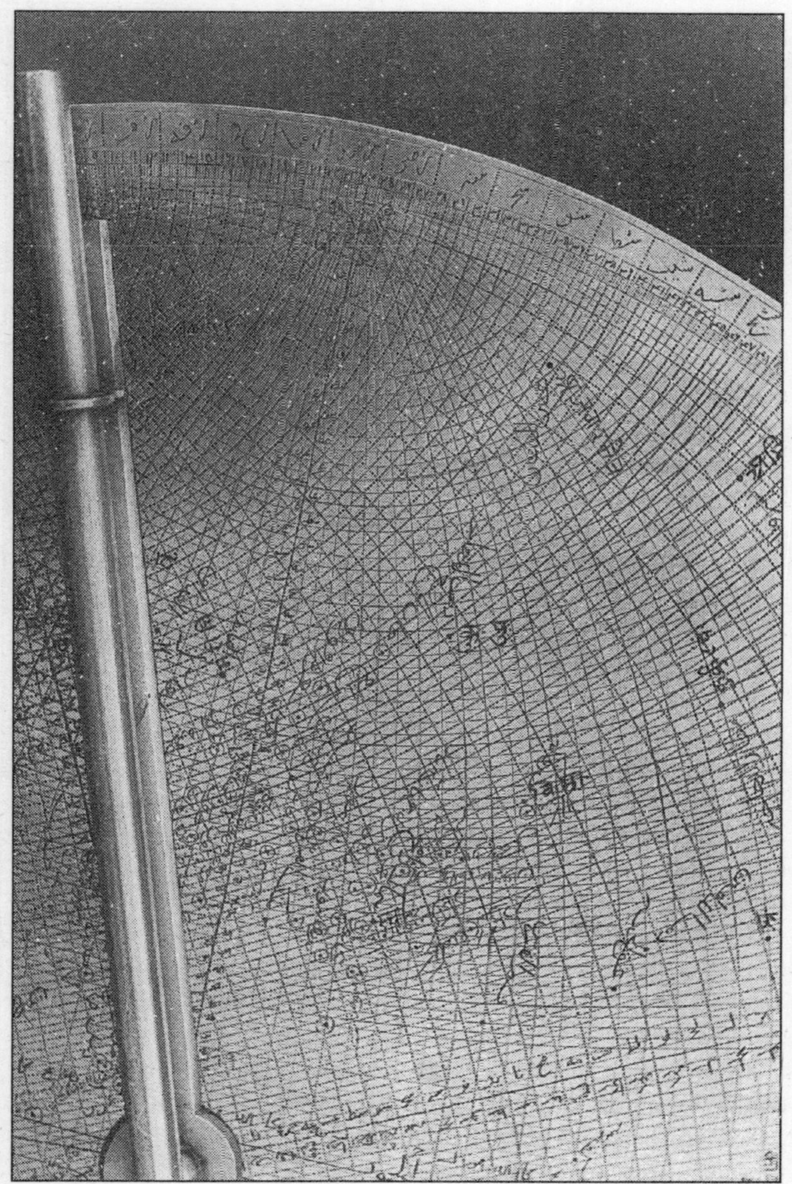

Fig. 11.2. Front of the astrolabe. Detail from the upper left quadrant.

geographical purposes. Treating the celestial equator and the polar axis as the zero degree latitude and longitude respectively, a number of town names were inscribed at appropriate distances. These include Ḥalab, Ṭūs, Kābul, Jahānābad (i.e. Delhi) and Lahore.[22]

On the back there are many scales and tables arranged in annular circles on the rim (see Fig. 11.3). Reading from the outer circumference to the centre, the annuli contain the following scales and tables. (i) The periphery of the two upper quadrants is graduated for every three degrees, numbered in *abjad*, then in single degrees numbered in Arabic numerals from 1 to 90, and then in one-sixths. The periphery of the two lower quadrants contains shadow scales, the 12-scale on the left and the 7-scale on the right. (ii) The next complete annulus contains the signs of the zodiac, divided in the succeeding circles into 3°, 1° and 10° respectively. (iii) Circle of lunar mansions. (iv-ix) Planets, arranged respectively twelve, nine, . . . to each sign, with varying graduations. (x) Shadow scales. (xi) Names of European months with a scale showing the days of each month. Here the first point of Aries is placed at 10 March, which is a correct value for the year AD 1680-81, when the instrument was made.

In the central part, the quadrant on the lower right contains a sine graph. The other three quadrants contain orthogonal projections of the great circles inclined to the meridian. Strangely enough the "Circle of the Moon", considered to be a special invention of al-Zarqālluh for finding the lunar distances,[23] is not engraved on the back of this astrolabe; nor is the circle mentioned in the Sanskrit manual on the Zarqālī astrolabe to be discussed below.

There are not many Zarqālī astrolabes extant today and none of these is from the East. Again none of the surviving ones is comparable to the magnificent creation of Ḍiyā' al-Dīn Muḥammad. While this piece at Jaipur is thus very unique, its subsequent history is also of great interest for the transmission of ideas. Within half a century of its manufacture, it was acquired by the astronomer-king Sawai Jai Singh of Jaipur (1688-1743), who is well known for the huge astronomical instruments in masonry which he erected at five principal towns of northern India, viz., Delhi, Jaipur, Varanasi,

22. Cf. Kaye, pp. 28-29.
23. Cf. Puig (v).

Fig. 11.3. Back view of the astrolabe.

Ujjain and Mathura.[24] He also collected some exquisitely crafted Mughal astrolabes, including the Zarqālī astrolabe which his court astronomers avidly studied. Under Jai Singh's orders, a Sanskrit / Rājasthānī name was engraved in Devanāgarī script at the back[25] of the crown of the Zarqālī astrolabe thus: *yantra jarakālī sarvadeśī*, literally "instrument Zarqālī universal". The first letters of the Sanskrit names of the zodiacal signs were also engraved at appropriate places on both sides of the ecliptic. Besides, a number of new star positions were marked with simple dots and their Sanskrit names were inscribed. Also a copper plaque was attached to the astrolabe, on which all the functions of this universal instrument were enumerated in Rājasthānī (see Fig. 11.4). The plaque, measuring 150 × 110 mm, reads as follows:

> One instrument [called] Zarqālī. This instrument is for all latitudes. This instrument provides [i] the knowledge of [time in] *ghaṭīs* in the day and at night; [ii] knowledge of the true [longitudes of the] planets; [iii] knowledge of the longitudes of the fixed stars; [iv] knowledge of the earth-sine; [v] sine of the ascensional difference; [vi] half duration of daylight; [vii] ascensional difference; knowledge of [viii] the azimuth, [ix] amplitude and [x] the terrestrial latitude; [xi] knowledge of the four *bhāvas* beginning with the ascendant; [xii] knowledge of the gnomon shadow; [xiii] knowledge of the day sine, and all other kinds of knowledge. This instrument consists of a single plate.

More important still is that Jai Singh caused the composition of a Sanskrit manual entitled *Sarvadeśīya-Jarakālī-Yantra* on the construction and use of this instrument.[26] Though there are several

24. On Jai Singh's masonry instruments, see Garrett; Kaye; Virendra Nath Sharma.
25. That is why Kaye, p. 27, treats the side having the name as the front and the side with the Zarqālī projection as the reverse, and goes on to say: "The characteristic part of the instrument is engraved on the reverse".
26. There are three independent manuscripts of this text: (i) Maharaja Sawai Mansingh II Museum, Jaipur, Khas Muhar 5483; (ii) Trinity College, Cambridge, R. 15.139, ff. 1-8; (iii) Bhandarkar Oriental Reseach Institute, Poona, 557 of 1899/1915. Besides these three, there are some other manuscript copies, which seem to have got mixed up with the leaves of other texts composed about the same time at Jaipur and these were printed with those other texts. Thus our manual occurs abruptly in Ram Swarup Sharma, vol. 3, pp. 1252-1260; Caturveda, pp. 96-105.

Fig. 11.4. Back of the astrolabe. Detail showing the inscription on the crown, and the plaque.

manuals in Sanskrit on the astrolabe,[27] this is the first time that the universal astrolabe invented in the eleventh century in far-off al-Andalus by Ibn al-Zarqālluh found an echo in Sanskrit writings.

There are forty sections in this work. The first section explains the construction of the front of the astrolabe, while the second lays down how to prepare the reverse side.[28] These two sections together are designated as *paribhāṣā*, "definition." The remaining part is called *gaṇita-prakāra*, "method of computations" and deals with the solving of various problems, such as finding the true longitude of the sun from the declination by three different methods; the converse of the problem, viz., finding the declination when the solar longitude is known, also by three different methods; finding the half duration of the daylight for any day at any latitude; finding the current ascendant and from it determining the same for any location or for another year, and so on.[29]

What are the sources of this text? One source clearly is the actual astrolabe manufactured by Ḍiyā' al-Dīn. Its presence at Jaipur, the similarity in the Sanskrit name engraved on its crown and the title of the manual, the near identity of the contents of the plaque and the manual strengthen the hypothesis that the instrument inspired the composition of the Sanskrit text.

Besides the actual instrument, literary documents may also have been the source for the Sanskrit manual. Ibn al-Zarqālluh himself is said to have composed three versions of his account, respectively in sixty, eighty and one hundred chapters.[30] Did any of these reach Jai Singh's court? If so, the Sanskrit manual falls into the category of texts Jai Singh got translated from Arabic and Persian sources.

27. For the various Sanskrit texts on the astrolabe, see Sarma (ii), pp. 238-241.
28. While explaining the construction of the instrument, the manual lays down that there should be an alidade with sighting vanes, and also a regula with a transversal cursor. This must indeed have been the case with Ḍiyā' al-Dīn's instrument when it was acquired by Jai Singh in the 1720s. The movable parts may have been lost after Jai Singh's time.
29. The titles of the forty sections will be given in the Appendix.
30. Cf. King, p. 253; Turner, pp. 155-156.

As I have shown elsewhere,[31] this process of translation proceeded in three separate stages. First a Muslim astronomer explained the Arabic/Persian text, sentence by sentence, to his Hindu counterpart in the local language. The latter immediately rendered each sentence into Sanskrit. In the third stage, one of the senior Hindu astronomers polished the translation, composed a preamble and a colophon, and signed it as his work. Our manual appears to be the product of the second stage. Only in one manuscript copy, the style has been polished, but no author put his signature on it.[32] Therefore it is clearly a translation or an adaptation of some Islamic text. A comparison then is essential of this Sanskrit text with Ibn al-Zarqālluh's three versions.

The universal astrolabe was discussed also in another contemporary work called *Yantrarājaracanā* which is attributed to Jai Singh himself. This work deals with the construction and use of the common northern astrolabe, but towards the end there is a section entitled "The method of observation with the universal astrolabe".[33] The problems solved in this section correspond to sections 3-16, 18 of the *Sarvadeśīya Jarakālī-Yantra.*

Until recently it was thought that Ḍiyā' al-Dīn's was the only universal astrolabe to have been produced in India and perhaps the only one of its size. In September 1992, Christie's of London offered for auction a yet larger universal astrolabe.[34] It has a diameter of 920 mm and is perhaps the largest known astrolabe in the world. The grid, however, is said to be of the *Shakkāziyya* type. The names of the zodiacal signs are engraved both in Arabic and Sanskrit, so also the arguments on the various legends. Other legends and numbers are written in Sanskrit only. On the reverse, close to the centre, there is a table of astrological "faces" and "limits", which is likewise written only in Sanskrit. There does not appear to be any maker's signature, nor any mention of the date or place of manufacture. Is it once again a product of the Lahore astrolabist family, on which Sanskrit legends were added later on,

31. Cf. Sarma (i), pp. 6-7; Sarma (vi).
32. MS Poona. Pingree, p. 319, attributes the translation to Nayanasukha.
33. Kedāranātha, pp. 13-17.
34. Christie's Auction Catalogue, item 119, pp. 48-49.

or is it a bilingual astrolabe manufactured wholly at Jai Singh's court? Whatever be the case, it is interesting that the most spectacular specimens of the *Ṣafīḥa Zarqāliyya* and the *Ṣafīḥa Shakkāziyya* should come from the fringes of Eastern Islam.

BIBLIOGRAPHY

Brieux, Alain & Maddison, Francis, *Répertoire des Facteurs d'astrolabes, et leurs oeuvres.* (In preparation).

Caturveda, Muralīdhara (ed.), *Siddhāntasamrāṭ Jagannāthasamrāḍvira-citaḥ.* Sagar, 1976.

Christie's Auction Catalogue, 24 September 1992, *Engineering and scientific Works of Art and Instruments, including Instruments from the Collection of the Late Saul Moskowitz.* Christie's, Manson and Woods Ltd. London, 1992.

Garrett, A. ff. & Guleri, Chandradhar, *The Jaipur Observatory and its Builder.* Allahabad, 1902.

Gibbs, Sharon L., Henderson, Janice A., Price, Derek de Solla, *A Computerized Checklist of Astrolabes.* Yale University. New Haven, 1973.

Gunther, Robert E., *The Astrolabes of the World.* Oxford, 1932.

Hartner, Willy, "Asṭurlāb", in idem, *Oriens-Occidens.* Hildesheim, 1968, pp. 312-318.

Kaye, G.R., *The Astronomical Observatories of Jai Singh*, Calcutta, 1918.

Kedāranātha Jyotirvid (ed.), *Śrīmanmahārājādhirāja-Jayasiṃhadeva-kāritā Yantrarājaracanā.* Jaipur, 1953.

King, David A., "On the Early History of the Universal Astrolabe in Islamic Astronomy, and the Origin of the term 'Shakkāziya' in Medieval Scientific Arabic" in idem, *Islamic Astronomical Instruments.* London, 1987. Article no. VII.

Pingree, David, "Indian and Islamic Astronomy at Jayasiṃha's Court" in: David A. King & George Saliba (eds.), *From Deferent to Equant: A Volume of Studies in the History of Science in the Ancient and Medieval Near East in Honor of E.S. Kennedy.* New York, 1987, pp. 313-328.

Puig, Roser (i), "Concerning the Ṣafīḥa Shakkāziyya," *Zeitschrift für die Geschichte der Arabisch-Islamischen Wissenschaften*, 2 (1985), pp. 123-139.

Puig, Roser (ii), *Al-Šakkāziyya. Ibn al-Naqqaš al-Zarqālluh. Edición, traducción y estudio.* Barcelona, 1986.

Puig, Roser (iii), "La proyección ortográfica en el Libro de la Açafeha alfonsí", *De Astronomia Alphonsi Regis.* Barcelona, 1987, pp. 125-138.

Puig, Roser (iv), *Los Tratados de Construcción y Uso de la Azafea de Azarquiel.* Madrid, 1987.

Puig, Roser (v), "Al-Zarqālluh's Graphical Method for Finding Lunar Distances", *Centaurus*, 32 (1989), pp. 294-309.

Raikva, K.K. (ed.), *The Yantrarāja of Mahendra Sūri with the Commentary by Malayendu Sūri.* Bombay, 1936.

Sarma, Sreeramula Rajeswara (i) (ed. & tr.), *Yantraprakāra of Sawai Jai Singh.* Supplement to *Studies in History of Medicine and Science*, 10-11 (1986-87).

Sarma, Sreeramula Rajeswara (ii), "Astronomical Instruments in Mughal Miniatures," *Studien zur Indologie und Iranistik*, Hamburg, 16-17 (1992), 235-276; reprinted in this Volume, pp. 76-121.

Sarma, Sreeramula Rajeswara (iii), "From al-Kura to Bhagola: On the Dissemination of the Celestial Globe in India," *Studies in History of Medicine and Science*, 13.1 (1994), 69-85; reprinted in this Volume, pp. 275-93.

Sarma, Sreeramula Rajeswara (iv), "The Lahore Family of Astrolabists and their Ouvrage," *Studies in History of Medicine and Science*, 13.2 (1994), 205-224; reprinted in this Volume, pp. 199-222.

Sarma, Sreeramula Rajeswara (v), "Portable Instruments at the Jaipur Observatory" to appear in the *Proceedings of the International Symposium on Indian and Other Asiatic Astronomies,* Hyderabad-Jaipur, 1991.

Sarma, Sreeramula Rajeswara (vi), "Translation of Scientific Texts under Sawai Jai Singh", *Sri Venkateswara University Oriental Journal*, 41 (1998), 67-87.

Sarma, Sreeramula Rajeswara (vii), "Sulṭān, Sūri and the Astrolabe," *Indian Journal of History of Science*, 35.2 (2000), 129-147; reprinted in this Volume, pp. 179-98.

Savage-Smith, Emilie, *Islamicate Celestial Globes: Their History, Construction and Use.* Washington D.C., 1985.

Sharma, Ram Swarup (ed.), *Samrāḍ Jagannātha-viracita Samrāṭ Siddhānta (Siddhānta-sāra Kaustubha).* New Delhi, 1967-69.

Sharma, Virendra Nath, *Sawai Jai Singh and his Astronomy*, Delhi, 1995.

Turner, A.J., *Astrolabes, Astrolabe-related Instruments.* The Time Museum: Catalogue of Collections, vol. I, part 1. Rockford, 1985.

APPENDIX

The Forty Sections of the *Sarvadeśīya Jarakālī-Yantra.*

1. Method of preparing the first side.
2. Method of drawing lines on the second side.
3. Finding the true longitude of the sun from the declination.
4. Another method.
5. Yet another method.
6. Finding the declination when the sun's longitude is known.
7. Another method.
8. Yet another method.
9. Finding the ascensional difference and the half duration of daylight.
10. Determining the time till/from noon (*natakāla*) and the time from sunrise or till sunset (*unnatakāla*).
11. Finding the altitude from the time till/from noon.
12. Finding the amplitude.
13. Determining the right ascensions at the equator.
14. Determining the ascendant from the right ascensions at the equator.
15. Determining the oblique ascensions.
16. Determining the ascendant from the oblique ascensions.
17. Determining the first visibility correction.
18. Determining the position of a planet or of a fixed star on the instrument.
19. Determining the ascensional difference, amplitude, half daylight of planets and fixed stars.
20. Determining the second visibility correction.
21. Finding the longitudinal difference of a planet that has moved from a given position.
22. Determining the altitudes of the planets and fixed stars.
23. Determining the true declination and the altitude from the time till / from noon.
24. Determining the time elapsed at night from the altitudes of the planets and fixed stars.
25. Determining the altitudes of the planets and fixed stars from the time elapsed at night.
26. Determining the ascendant.
27. Determining the time elapsed in the day or at night from the culmination.
28. Determining the ascendant from the culmination.
29. When the ascendant and the culmination are known, determining the altitude for each degree of the ecliptic.
30. Determining the culmination from the ascendant.

31. Determining the degrees of the ecliptic from the altitude.
32. Determining the time difference between a planet and the meridian and, from that difference, determining the azimuth and altitude.
33. Determining the azimuth from the altitude.
34. Determining the time till/from noon and the declination from the azimuth.
35. Finding the direction of another city in relation to a given city.
36. Finding the arc intercepted by two planets.
37. If the time till/from noon in one city is known, finding the same for another city.
38. Finding the shadow from the altitude and vice versa.
39. When the ascendant in one year is known, finding it for other years.
40. Finding the ascendant at another latitude from the ascendant at one's own latitude.

12

Yantrarāja: The Astrolabe in Sanskrit

0.1 The astrolabe is a highly sophisticated astronomical instrument of the pre-modern times. In the Middle Ages, it enjoyed a reputation comparable to that reserved today for the personal computer. It is a versatile observational and computational instrument. As an observational instrument, it was employed for measuring the altitudes of the heavenly bodies and also for measuring the heights and distances in land survey. As a computational device, it can be made to simulate the motion of the heavens at any given locality and time. It can also be used as an analog computer for solving numerous problems in spherical trigonometry. Therefore, the astrolabe was rightly termed as the "universe within one's palm."[1]

The invention of the astrolabe is usually attributed to Hipparchus of the second century BC. But there is no firm evidence to support

First published in: *Indian Journal of History of Science*, 34.2 (1999) 145-58.

Abbreviations used:
CESS = Pingree, David. *Census of the Exact Sciences in Sanskrit*, Series A, vol. 1: 1970; vol. 2: 1971; vol. 3: 1976; vol. 4: 1981; vol. 5: 1994; all published from Philadelphia.
IJHS = *Indian Journal of History of Sciences*, New Delhi.
SHMS = *Studies in History of Medicine and Science*, New Delhi.

1. The literature on the history, construction and principles of the astrolabe is quite vast. The best introduction is Hartner, Willy, The Principles and Use of the Astrolabe, in: idem, *Oriens-Occidens*, [vol. I], Hildesheim 1985, pp. 287-311. For its history in India, see Sarma, Sreeramula Rajeswara, Astronomical Instruments in Mughal Miniatures, *Studien zur Indologie und Iranistik*, 1992, 16, pp. 235-276; reprinted in this Volume, pp. 76-121; idem, Indian Astronomical and Time-Measuring Instruments: A Catalogue in Preparation, IJHS, 1994, 29.4, pp. 507-528; reprinted in this Volume, pp. 19-46; and Ohashi, Yukio, Early History of the Astrolabe in India, IJHS, 1997, 32.3, pp. 199-295.

this view. It is however certain that the instrument was well known to the Greeks before the beginning of the Christian era. Even so, the astrolabe reached its highest perfection and popularity in the Islamic World. Muslim astronomers contributed several treatises on the science of the astrolabe. They also transmitted the knowledge of the astrolabe to Europe in the west and to India in the east.

0.2 It is not known when exactly the astrolabe reached India. Al-Bīrūnī claims, in his *Indica,* to have composed a manual on the astrolabe in Sanskrit verse.[2] Whether true or not,[3] there is no denying the fact that he discussed the astrolabe and its variants in several of his works.[4] Therefore it is entirely probable that he brought the astrolabe with him and taught its working principles to the Hindu *jyotiṣīs* at Multan in the first quarter of the eleventh century. With the establishment of the Sultanate in Delhi in the next century, Muslim scholars began to migrate in great numbers to Delhi. Some of these brought astrolabes and employed them for teaching astronomy or for astrological purposes. Since astronomy formed a regular part of Muslim education, ability to use the astrolabe was expected of every good scholar. A few thirteenth century astrolabes from the Middle East are still extant in Indian collections, which must have been brought here by the immigrant scholars.

In the latter half of the fourteenth century, Fīrūz Shāh Tughluq promoted the study of the astrolabe.[5] He not only got several astrolabes manufactured at Delhi—probably for the first time in India—he also encouraged the composition of manuals on the construction and use of the astrolabe both in Persian and in Sanskrit. During the next five hundred years up to the end of the nineteenth century, a large number of astrolabes were manufactured in India with Arabic/Persian legends. These are generally classified

2. Cf. Sachau, Edward (tr), *Alberuni's India,* first Indian reprint: New Delhi, 1964, vol. I, p. 137.
3. Doubts were cast on his knowledge of Sanskrit; cf. among others, Pingree, David, Al-Bīrūnī's Knowledge of Sanskrit Astronomical Texts, in: Chelkovski, Peter J. (ed), *The Scholar and the Saint*, New York, 1975, pp. 67-81.
4. Cf. Sarma, Sreeramula Rajeswara, "Sulṭān, Sūri and the Astrolabe," IJHS, 35.2 (2000) 129-147; reprinted in this Volume, pp. 179-98.
5. On Fīrūz's interest in the astrolabe, see ibid.

as Indo-Persian astrolabes. I have discussed this class of astrolabes elsewhere in greater detail.[6]

0.3 In this paper, I wish to examine the response of Hindu and Jaina *jyotiṣīs* to this remarkable instrument imported from the Islamic World. Their response was twofold: first, several manuals were composed in Sanskrit on the construction and use of the astrolabe. Secondly, this activity of composition was accompanied by the manufacture of astrolabes on which the legends were engraved in Sanskrit language and in Devanāgarī script. I classify this group as "Sanskrit astrolabes".

Between 1370 and 1870, at least fifteen manuals were composed in Sanskrit which discuss the astrolabe either exclusively or as one of the several astronomical instruments. The Sanskrit astrolabes that still survive do not go that far back. The earliest extant Sanskrit astrolabe was crafted in Gujarat in 1607 and the latest that I saw was manufactured in Rajasthan most probably in 1902. Between these two lie some seventy astrolabes which are either extant or for which there are records.

0.4 This impressive number of Sanskrit manuals and Sanskrit astrolabes produced during the half millennium from 1370 to the end of the nineteenth century makes it abundantly clear that the Indian *jyotiṣīs* were greatly impressed by this versatile instrument. In fact, no other foreign idea or object enthused Indian pundits in pre-modern times as much as the astrolabe seems to have done.

0.5 In the following pages, I present first an overview of the literature on the astrolabe in Sanskrit[7] and then of the extant specimens of the Sanskrit astrolabes. Some of these Sanskrit texts and several of the signed Sanskrit astrolabes will be taken up for detailed study in my forthcoming publications, as part of an ongoing INSA project on "A Descriptive Catalogue of Indian Astronomical and Time-Measuring Instruments".

1.1 Of the significant contribution made by the Jainas to the intellectual history of India, an important but not so well explored aspect is their role as mediators between the Islamic and Sanskritic

6. Sarma, Sreeramula Rajeswara, The Lahore Family of Astrolabists and their Ouvrage, SHMS, 1994, 13.2, pp. 205-224; reprinted in this Volume, pp. 199-222.
7. For the sake of clarity, a list of these texts is given in the Appendix.

traditions of learning.[8] After the establishment of the Sultanate of Delhi, a number of Jaina monks and laymen cultivated good relations with the Delhi court. One of these was Mahendra Sūri, the author of the first ever manual on the astrolabe in Sanskrit.[9] The credit for sponsoring this manual, as has been stated above, goes to Fīrūz Shāh Tughluq. Under his patronage, Mahendra Sūri produced the manual in 1370—at about the same time as Geoffrey Chaucer wrote his *Treatise on the Astrolabe* in English. The Sūri was so impressed by the versatile functions of the astrolabe that he called it *yantrarāja,* "the king of astronomical instruments", and it is under this name that the astrolabe came to be known in Sanskrit since then. Accordingly his manual bears the title *Yantrarājāgama* or simply *Yantrarāja.*[10] His pupil Malayendu Sūri[11] wrote a commentary on this work in about 1382. From Malayendu we learn that his teacher Mahendra Sūri was the foremost astronomer at Fīrūz's court (*śrīpīrojaśakendra-sarvagaṇaka-praṣṭho mahendra-prabhuḥ*).[12] There are two other commentaries on this *Yantrarāja*, by Gopīrāja written in 1540,[13] and by Yajñeśvara in 1842.[14]

The text is divided into five chapters. The first chapter *Gaṇitādhyāya* provides various trigonometrical parameters needed for the construction of the astrolabe. The second chapter called *Yantraghaṭanādhyāya* enumerates the different parts of the astrolabe. The construction of the common northern astrolabe (*saumya-yantra*) and other variants is described in the third chapter named *Yantraracanādhyāya*, while the next *Yantraśodhanādhyāya* explains the method of verifying whether the astrolabe is properly constructed or not. The final and fifth chapter, entitled *Yantravicāraṇā-*

8. I have argued this point in Sarma, Sreeramula Rajeswara, Sanskrit Manuals for Learning Persian, in: Safavi, Azarmi Dukht (ed), *Adab Shenasi*, Aligarh, 1996, pp. 1-12.
9. Cf. CESS, A-4, 393-395; A-5, 296.
10. Raikva, Kṛṣnaśaṃkara Keśavarāma (ed), *Yantrarāja of Mahendra Sūri together with the commentary of Malayendu Sūri and Yantraśiromaṇi of Viśrāma,* Bombay, 1936.
11. Cf. CESS, A-4, 363-364; A-5, 282-3.
12. In a common stanza that occurs at the conclusion of his commentary on each of the five chapters.
13. Cf. CESS, A-2, 133.
14. Cf. ibid, A-5, 318-319. I have used MS BORI 556 of 1899-1915.

dhyāya discusses the use of the astrolabe as an observational and computational instrument and dwells on the various problems in astronomy and spherical trigonometry that can be solved by means of the astrolabe.[15]

At the beginning of his work, Mahendra says that the Muslims (*yavana*) have written many treatises on the astrolabe. Having extracted their essence, just as one extracts nectar after churning the milky ocean, he is presenting this work in Sanskrit.[16] We do not know exactly what these Arabic/Persian sources were which Mahendra Sūri consulted, but they certainly must have included al-Bīrūnī's various Arabic writings on the astrolabe and also al-Marrākushī's thirteenth century treatise on the astronomical instruments.[17]

In the same year when Mahendra Sūri wrote the *Yantrarāja*, there appeared also the *Sīrat-i Fīrūz Shāhī*, an anonymous chronicle of the reign of Fīrūz Shāh Tughluq.[18] In one of its chapters, the chronicle describes in great detail Fīrūz's interest in the astrolabe and the various astrolabes he got manufactured at Delhi. There is a great correspondence between this account and that in Mahendra's Sūri's *Yantrarāja.* This fact suggests that Mahendra Sūri and the unnamed Muslim astronomers at Fīrūz's court were working in close cooperation.

Therefore it is not surprising that, following the Persian *Sīrat,* the Sanskrit *Yantrarāja* also discusses two lesser known variants of the astrolabe. The most common variety of astrolabe is the so-called northern astrolabe (Arabic: *asṭurlāb shumālī*; Sanskrit: *saumya-yantra*). Here the rete is projected from the south celestial pole and displays a map of the northern celestial hemisphere,

15. Cf. also Ohashi, op. cit., pp. 211-216.
16. *Yantrarāja*, 1.3.
17. Professor David Pingree informs me that the table of the day-sines in Mahendra Sūri's *Yantrarāja* (pp. 11-15) is derived from al-Marrākushī. For the latter, cf. Śedillot, Jean-Jacques & Louis-Amélie, *Traité des instruments astronomiques des Arabs*, Paris, 1834; reprint: Frankfurt, 1984, pp. 349-350.
18. The unique copy of this chronicle is preserved in the Khuda Bakhsh Oriental Public Library, Patna. I had consulted the unpublished English translation by Professor Syed Hasan Askari. Cf. also Sarma, Sulṭān, Sūri and the Astrolabe (n. 4).

extending up to the Tropic of Capricorn. Therefore, with this type of astrolabe, the orientation is done at night by means of stars situated to the north of the Tropic of Capricorn. As against this, the southern astrolabe (Arabic: *asṭurlāb janūbī*; Sanskrit: *yāmya-yantra*) displays the southern celestial hemisphere and here orientation is done by means of those stars that lie to the south of the Tropic of Capricorn and are visible in the northern temperate zone of the earth. The third variant combines the features of these two and is called *asṭurlāb shumālī wa janūbī* (Sanskrit: *miśra-yantra*). For constructing the north-south astrolabe, Malayendu provides in his commentary elaborate tables of eccentricities (*kendra*) and radii (*vyāsārdha*), in northern and southern hemispheres, at six degree intervals of altitudes, for six different localities.[19]

1.2 At the beginning of the fifteenth century, Padmanābha[20] devoted the first chapter *Yantrarājādhikāra* of his *Yantrakiraṇāvalī* to the astrolabe. Strangely enough, Padmanābha does not discuss in this chapter the common northern astrolabe but the rather unusual southern astrolabe.[21]

1.3 In 1428, Rāmacandra Vājapeyin[22] discussed the astrolabe quite extensively in his *Yantraprakāśa*[23] which he composed at Pātrapuñjanagara, near modern Lucknow in U.P. The *Yantraprakāśa* describes the construction and use of some thirty-five astronomical instruments—perhaps the largest number ever dealt with in a Sanskrit work. The major part of this work is devoted to the

19. Ibid, pp. 19-25. These localities are Tilaṅga (latitude 18°), Tryambaka (21°), Aṇahillapattana (24°), Tīrabhukti (27°), Dillī (28°39′) and Nepālapura (31°).
20. Cf. CESS, A-4, 170-172; A-5, 205.
21. It is unusual because, of the three thousand and odd Islamic, European and Indian astrolabes that are extant today in different parts of the world, there are hardly a dozen southern astrolabes. In his Early History of the Astrolabe in India (n. 1), Dr. Yukio Ohashi offers a fine edition, translation and commentary of Padmanābha's *Yantra-rājādhikāra*.
22. Cf. CESS, A-5, 467-479.
23. Together with an auto-commentary, available in MS G-1363 of the Asiatic Society, Calcutta and MS 975/1886-92 of the Bhandarkar Oriental Research Institute, Poona.

astrolabe, which is called here *Sulabhā,* another significant name meaning that with this instrument several types of measurements become easy. Rāmacandra expressly declares that if one knows the science of the astrolabe well, the entire universe will become comprehensible like the myrobalan on one's palm (*yasmin karāmalakavad vidite viditaṃ bhaved viśvam*).[24]

Though Rāmacandra does not mention Mahendra Sūri by name, his familiarity with the latter's *Yantrarāja* and also with Malayendu's commentary is clearly discernible in his work. Rāmacandra appears to have a genuine interest also in the practical aspect of instrument making. He discusses not only the theory of the astrolabe but dwells also on the various tools and devices needed in its manufacture.

1.4 In this century, the science of the astrolabe seems to have reached Kerala as well. In his commentary entitled *Siddhāntadīpikā* composed in 1432 on the *Mahābhāskarīya*, Parameśvara[25] uses the shadow squares at the back of the astrolabe to measure the altitude of the eclipsed body. Here Parameśvara gives the altitudes in terms of the shadows of a gnomon of 6 feet (*padabhā*).[26] This is rather unusual because the Sanskrit astrolabes usually contain shadow squares for only gnomons of 7 digits (*saptāṅgulaśaṅkucchāyā*) and of 12 digits (*dvādaśāṅgulaśaṅkucchāyā*) as against the shadow squares respectively for 7 feet and for 12 digits in the Islamic astrolabes. Thus it is likely that Parameśvara's knowledge of the astrolabe is based on a tradition that is different from the one prevailing in western and northern India.

1.5 We have noted the role of the Jaina monk Mahendra Sūri as mediator between the Islamic and Sanskritic tradition of learning. About the end of the fifteenth century, another Jaina monk, Muni Megharatna, pupil of Vinayasundara of Vaṭagaccha, wrote the *Usturalāvayantra* in 38 stanzas, with many Arabic-Persian technical

24. *Yantraprakāśa* 1.9.
25. Cf. CESS, A-4, 187-192.
26. Cf. *Mahābhāskarīya of Bhāskarācārya with the Bhāṣya of Govindasvāmin and the Super-commentary Siddhāntadīpikā of Parameśvara,* ed. T.S. Kuppanna Sastri, Madras, 1957, pp. 330-331. Professor David Pingree kindly drew my attention to this passage.

terms. The unique manuscript copy at the Anup Sanskrit Library of Bikaner contains also commentaries in Sanskrit and Rajasthani.[27]

1.6 Three seventeenth century texts discuss the astrolabe along with other instruments. In 1615 Viśrāma[28] of Jambūsara devoted the third chapter of his *Yantraśiromaṇi* to the astrolabe.[29]

In 1621 Nṛsiṃha Daivajña[30] of Kāśī discussed the astrolabe in his *Vāsanāvārttika* on the *Siddhāntaśiromaṇi,* quoting extensively from the works of Mahendra and Rāmacandra.[31]

In 1639 Nityānanda[32] devoted the *Yantrādhyāya* of his *Sarvasiddhāntarāja* mainly to the astrolabe. Like Mahendra Sūri, Nityānanda also divides his discussion of the astrolabe in five chapters and names them in a similar fashion, viz., *Gaṇitādhyāya, Ghaṭanādhyāya, Racanādhyāya, Śodhanādhyāya* and *Yantranirīkṣaṇādhyāya.*[33]

1.7 In the early eighteenth century, the study of the astrolabe received a great impetus under Sawai Jai Singh.[34] Jai Singh immediately saw the advantages of the astrolabe as a teaching tool, although for observational purposes he preferred large scale instruments in masonry. He collected a number of exquisitely crafted Mughal, i.e. Indo-Persian astrolabes. He also established a manufactory of Sanskrit astrolabes.[35] He himself composed or

27. Agar Chand Nahata, "Ustaralāva Yantra sambandhī ek Mahattvapūrṇa Jaina Grantha," *Jaina-siddhānta-bhāskara*, vol. 18.2, pp. 119-128.
28. Cf. CESS, A-5, 658.
29. Raikva, Kṛṣṇaśaṃkara Keśavarāma (ed), *Yantrarāja of Mahendra Sūri together with the Commentary of Malayendu Sūri, and Yantraśiromaṇi of Viśrāma*, Bombay, 1936, pp. 102-113.
30. Cf. CESS, A-3, 204-206.
31. Caturvedi, Murali Dhara (ed), *Siddhāntaśiromaṇi of Bhāskarācārya with the Vārttika of Nṛsiṃha Daivajña*, Varanasi, 1981, pp. 445-458.
32. Cf. CESS, A-3, 173-174; A-4, 141; A-5, 184.
33. Cf. Velankar, H.D., *A Descriptive Catalogue of the Sanskrit and Prakrit Manuscripts in the Collection of the Asiatic Society of Bombay*, 2nd edn, ed. V.M. Kulkarni & Devangana Desai, Mumbai, 1998, pp. 88-89, MS 264: *Sarvasiddhāntarāja* (*Yantradhyāya*) of Nityānanda.
34. Cf. CESS, A-3, 63-4; A-5, 117-118.
35. Some of these portable instruments, either collected by Jai Singh or manufactured under his orders, still survive in Jaipur. The Department of Archaeology and Museums, Government of Rajasthan, has recently constructed a separate building in the premises of the Jantar Mantar Observatory, Jaipur, to display these instruments.

caused to be composed four manuals on the astrolabe in Sanskrit. Like the Arabic and Persian scientific texts, all the four are written in prose. Thus under Sawai Jai Singh Sanskrit scientific writing was attempted for the first time in prose, abandoning the traditional verse form.

The first of these is entitled *Yantraprakāra*. It is a compilation of material on various astronomical instruments prepared during the early stages of Jai Singh's researches in astronomy from diverse sources including Ptolemy's *Almagest*. The eighth section of this work explains the use of the astrolabe rather briefly.[36] But a more detailed treatment of the astrolabe is available in the *Yantrarājaracanā,* attributed to Jai Singh. Here the astronomer-king makes an excellent presentation of the theory of the astrolabe. While Mahendra Sūri provides mathematical proofs for his propositions, Jai Singh offers geometrical proofs.[37]

One of the Mughal astrolabes collected by Jai Singh belongs to a special variety known as the "Universal Zarqālī Astrolabe." The original prototype was invented by Ibn al-Zarqālluh at Toledo in the eleventh century. The Mughal specimen acquired by Jai Singh was crafted by Dīyā' al-Dīn Muḥammad in 1681 at Delhi. It is a very large astrolabe measuring some 555 mm in diameter.[38] After acquiring this precious astrolabe, Jai Singh caused the composition of a Sanskrit manual entitled *Sarvadeśīyā-Jarakālī-Yantra* on the construction and use of this instrument.[39]

36. Sarma, Sreeramula Rajeswara (ed & tr), *Yantraprakāra of Sawai Jai Singh,* SHMS, 1986-87, 10-11, supplement; on the astrolabe, cf. pp. 20-21, 61-63.
37. Kedāranātha Jyotirvid (ed), *Yantrarājaracanā of Jayasiṃhadeva & Yantraprabhā of Śrīnātha,* Jaipur 1953.
38. Cf. Sarma, Sreeramula Rajeswara, The Ṣafīha Zarqāliyya in India, in: Casulleras, Josep & Samsó, Julio (ed), *From Baghdad to Barcelona: Studies in Islamic Exact Sciences in Honor of Prof. Juan Vernet,* Barcelona 1996, pp. 718-735; reprinted in this Volume, pp. 223-39; see also Kaye, G.R., *The Astronomical Observatories of Jai Singh,* Calcutta, 1918, pp. 27-30.
39. The text is inserted in the middle of the *Spaṣṭādhikāra* in: Caturveda, Muralīdhara (ed), *Siddhāntasamrāṭ, Jagannātha-Samrāḍ-viracitaḥ,* Sagar, 1976, pp. 96-105. I have also used the manuscripts at Maharaja Sawai Mansingh II Museum, Jaipur, Khas Muhar 5483; Trinity College, Cambridge, R. 15.139, ff. 1-8; BORI, Poona, No. 557 of 1899/1915.

In the Islamic World and also in India, Naṣīr al-Dīn Muḥammad al-Ṭūsī's (1201-1274) Persian manual on the astrolabe entitled *Risālat al-usṭurlāb* enjoyed great popularity. Since it consists of twenty chapters, it is commonly known as the *Bist Bāb* ("Twenty Chapters").[40] Under Jai Singh's orders, this work was translated into Sanskrit with the title *Yantrarāja-vicāra-viṃśādhyāyī.*[41] The Sarasvati Bhavan Library of the Varanasi Sanskrit University possesses an interesting manuscript which contains the Persian text of the *Bist Bāb* but in Devanāgarī script.[42] I am inclined to think that this transliteration was also produced at Jai Singh's court for some pundit who understood Persian but could not read the script.

1.8 Even after Jai Singh's time, the astrolabe continued to be discussed in several Sanskrit works. Śrīnatha Chagāṇī prepared a short metrical version in 29 verses of Jai Singh's *Yantrārajaracanā* with the title *Yantraprabhā.*[43] In 1772 Nandarāma[44] composed the *Yantrasāra* in which the astrolabe was discussed along with other instruments.[45]

Ten years later, in 1782, Mathurānātha Śukla[46] produced at Varanasi the *Yantrarāja-ghaṭanā,* also known as *Yantrarājakalpa.* Mathurānātha, who later became a teacher of astronomy in the Benares Sanskrit College founded by Jonathan Duncan, was a good scholar of Persian as well. In this work, Mathurānātha teaches the construction of all the three varieties of astrolabes—the northern, southern and the mixed—with the aid of detailed diagrams.[47] Finally

40. Cf. CESS, A-3, 145; A-4, 125.
41. Bhaṭṭācārya, Vibhūtibhūṣaṇa (ed), *Yantrarāja-vicāra-viṃśādhyāyī* by Nayanasukha Upādhyāya, Varanasi 1979.
42. MS. U 34568 catalogued under the title *Yantrarājaprayogaḥ.*
43. This text is published with Jai Singh's *Yantrarājaracanā* (see n. 37 above), on pp. 17-19.
44. Cf. CESS, A-3, 128-130; A-5, 156-158.
45. I have used MS BORI 851 of 1884-87 copied in 1802 and 504 of 1892-95 copied in 1830.
46. Cf. CESS, A-349-350.
47. MS, U 35245 of the Sanskrit University, Varanasi. In this manuscript, several of the diagrams are incomplete.

in the second half of the nineteenth century, Bāpudevaśāstrin[48] is reported to have prepared *Yantrarājopayogi-chedyaka* ("projections or drawings useful for the astrolabe"),[49] but no manuscript has come to light so far.

2.0 Mere manuals on the astrolabe do not serve much purpose unless there are actual specimens with which one can put the theory into practice. As early as 1790, one Reuben Burrow noticed a Sanskrit astrolabe brought to him by Dr. Mackinnon from Jaipur.[50] Soon Sanskrit astrolabes began to be collected by various employees of the East India Company, including H.H. Wilson, who later became the first Boden Professor of Sanskrit at the University of Oxford in 1832. Some of these astrolabes were studied by W.H. Morley in an appendix to his *Description of a Persian Astrolabe constructed for Shāh Husain Safawī*[51] in 1856. In this century, Kaye and others discussed Sanskrit astrolabes along with other instruments. Writing in 1985, Dr. Emilie Savage-Smith remarks in her exemplary study of the *Islamicate Celestial Globes* that "eight Sanskrit astrolabes are known to exist."[52] My investigation happily raises the number to about 70 for which records or references are available. Of these I have personally examined more than thirty.

48. Cf. CESS, A-4, 241-242; A-5, 232; cf. also, Sarma, Sreeramula Rajeswara, Sanskrit as Vehicle for Modern Science: Lancelot Wilkinson's Efforts in the 1830's, SHMS, 1995-96, [14], pp. 189-99.
49. Thus Śaṅkara Bālakṛṣṇa Dīkṣita in his *Bhāratīya Jyotiṣa*, tr. in Hindi by Śivanātha Jhārakhaṇḍī, 2nd edn, Lucknow 1963, p. 411.
50. Cf. Burrow, Reuben, A Proof that the Hindoos had the Binomial Theorem, *Asiatick Researches*, 1770, 2, pp. 486-497, esp. 488-489. Finding that the astrolabe from Jaipur exactly corresponds to the description given by Chaucer in his treatise and having just heard about the Vedas that they contain all knowledge, Burrow drew the fantastic conclusion that Chaucer's *Treatise on the Astrolabe* may have been translated from the Veda! I refrain from quoting this naive gentleman; far too many people are already looking for just this kind of endorsement from a westerner about the "scientific content" of the Veda.
51. Reprinted in Gunther, Robert T. *Astrolabes of the World*, Oxford 1932, vol. I, pp. 1-47. Appendix, No. 1, is on pp. 32-46.
52. *Islamicate Celestial Globes: Their History and Construction and Use*, Washington, DC, 1985, p. 304, n. 181.

2.1 It may be assumed that with Mahendra Sūri's composition of the first Sanskrit manual in 1370, the production of Sanskrit astrolabes also must have commenced. But they did not survive the vagaries of time and climate. Of those that are extant, the earliest known Sanskrit astrolabe is dated Saṃvat 1663, Śāka current 1528, Māgha vadi 1 Sunday, corresponding to 1 February 1607. The inscription on the astrolabe goes on to say that it was manufactured at Ahmedabad during the reign of Salīm Shāh, i.e., Jahāngīr, for the use of Dāmodara, son of Caṇḍīdāsa. It is a large piece with a diameter of 276 mm. There are six tympans (*akṣapatra*) calibrated for the latitudes of 12 different localities ranging from Bijapur (18°) in the south to Kashmir (35;20°) in the north. It is now in a private collection at Brussels. Doubts have been expressed about its authenticity. I have examined it very carefully and compared with three other Sanskrit astrolabes produced in the seventeenth century. I am convinced that it is genuine except for the crude star map (*bhapatra*) which is clearly a late replacement.[53]

2.2 Chronologically the next Sanskrit astrolabe is now at the Royal Scottish Museum of Edinburgh. It is dated Saṃvat 1700 Caitra kṛṣṇa 11, which corresponds to 2 April 1644. The inscription records that it was caused to be made for one Maṇirāma. Subsequently one Līlānātha Jyotirvid owned it. This fact is recorded in a different hand. The astrolabe measures 128 mm in diameter. It also contains six tympans for 12 different localities. This astrolabe is in an excellent state of preservation.

2.3 The next astrolabe for which we have records was manufactured on Friday, Śaka 1591 current Āsvayuja sudi 1, which corresponds to 12 October 1668, for Rāghavajit, son of Daivajña

53. This astrolabe was formerly at the Time Museum, Rockford, USA. For a detailed description, see Turner, A.J., *The Time Museum, Catalogue of the Collection,* vol. I, part 1: *Astrolabes, Astrolabe-related Instruments,* Rockford 1985, No. 15, pp. 120-123. In 1988, it was sold through the Christie's, South Kensington, London; cf. their auction catalogue *From the Time Museum, Time Measuring Instruments,* 14 April 1988, item no. 157, pp. 98-99. The present owner very kindly allowed me to study this astrolabe at Brussels in February 1996. I shall discuss this and the other seventeenth century Sanskrit astrolabes in a forthcoming paper.

Viśvanātha. In 1936, this astrolabe belonged to Kṛṣṇaśaṃkara Keśavarāma Raikva of Surat, who reproduced several drawings of this astrolabe and its various components in his edition of Mahendra Sūri's *Yantrarāja*[54] (Fig. 12.1). In his introduction to the *Yantrarāja,* Raikva mentions that he had in his collection also a manuscript of Mahendra Sūri's *Yantrarāja* which the same Rāghavajit had copied in Śaka 1590, i.e. just a year before the astrolabe was manufactured. Unfortunately, the present location of this historically important astrolabe is not known, nor of the manuscript.

2.4 The Pitt Rivers Museum of Oxford has one astrolabe which was produced on Tuesday, Saṃvat 1730 Kārtika sudi 6, corresponding to 17 October 1673, for the astrologer Indrajīka or Indrajī. With a diameter of 115 mm, it has five tympans on which projections are engraved for 8 different localities.[55]

2.5 As mentioned above, in the early eighteenth century Sawai Jai Singh established a *kārkhānā* for the manufacture of Sanskrit astrolabes. Here he caused the manufacture of some ornate astrolabes with multiple tympans, designed after the Mughal Indo-Persian astrolabes. He, however, popularised simple astrolabes with a single tympan calibrated to the latitude of Jaipur at 27° (see Fig. 12.2). It appears that a large number of such astrolabes were produced and distributed among the *jyotiṣīs* of his court so that they all became proficient in the science of the astrolabe. In the stores of Jai Singh's Observatory at Jaipur, I discovered a number of unfinished astrolabes belonging to this *kārkhānā.* After Jai Singh's death, this manufactory seems have shifted to Kūcamaṇa, a little to the west of Jaipur but on the same latitude. This place produced a number of interesting astrolabes.

2.6 Astrolabes were also crafted at other towns of Rajasthan, such as Bundi. Here was produced an enormous astrolabe measuring 662 mm in diameter which is now with the Science Museum of London. The astrolabist Śivalāla completed this on

54. See n. 10 above; one of these drawings is reproduced here as Figure 12.1.
55. A brief description of this astrolabe appeared in Gunther, R.T. *Early Science at Oxford*, Oxford, 1923, vol. II, pp. 279-280. See also idem, *The Astrolabes of the World*, Oxford 1932, vol. I, No. 79, p. 211, Fig. 110.

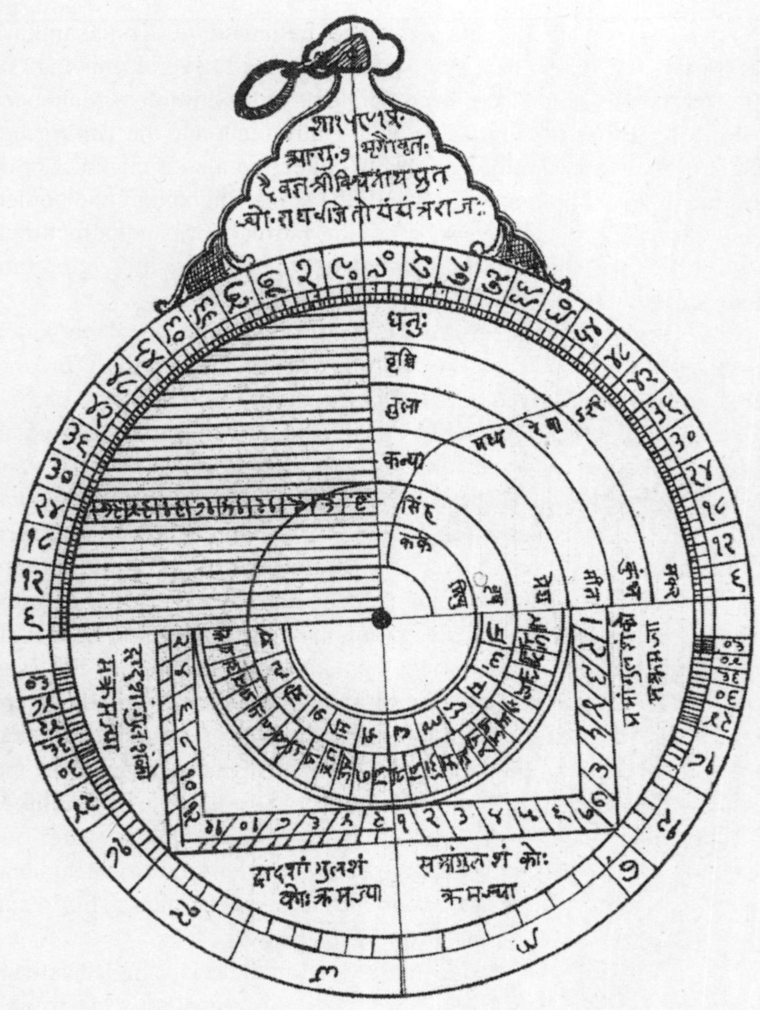

Fig. 12.1. Rāghavajit's Sanskrit astrolabe with five tympans, AD 1668. Back with the inscription engraved on the suspension bracket at the top. (From K.K. Raikva's edition of the *Yantrarāja*.)

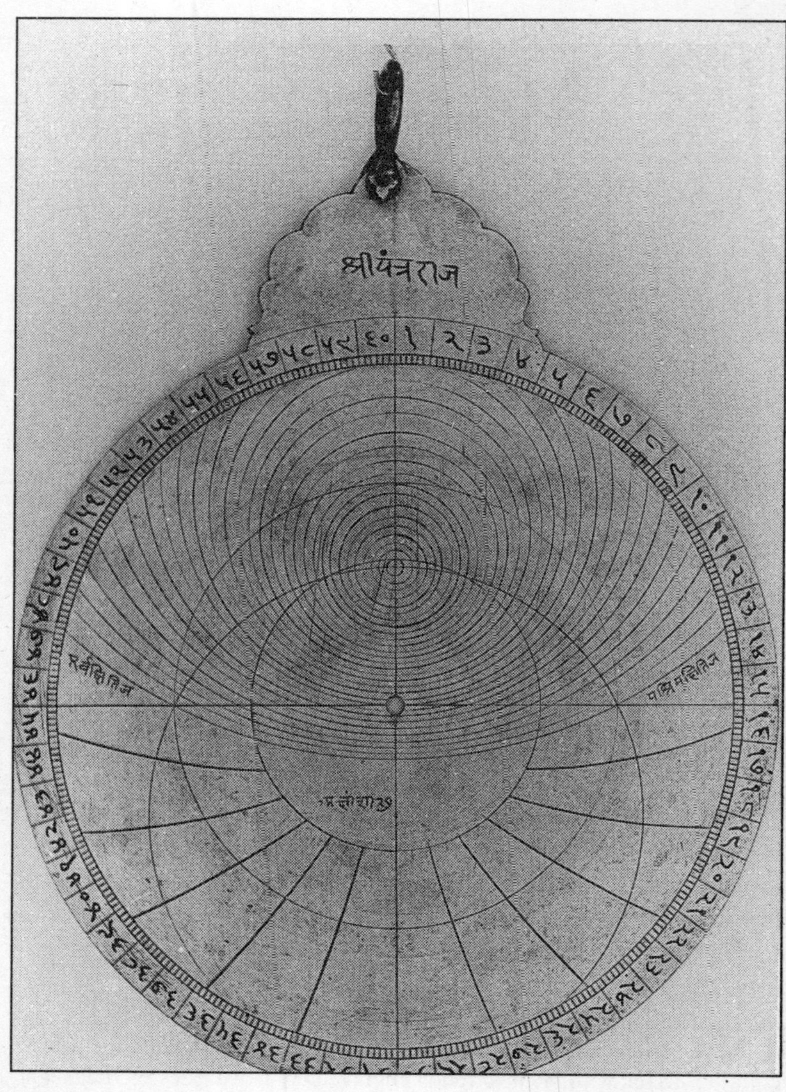

Fig. 12.2. A Sanskrit astrolabe with a single tympan calibrated for the latitude of 27° N. Note the name *Śrīyantrarāja* engraved on the suspension bracket at the top and the latitude *akṣāṃśa* 27 engraved in the middle. Present location unknown. Archive photo: Courtesy, The Museum of the History of Science, Oxford.

Fig. 12.3. Sanskrit astrolabe manufactured by Śivalāla in 1879 for the latitude of Bundi at 25;30° N. The entire fifth chapter of Mahendra Sūri's *Yantrarāja* is engraved in the two upper quadrants at the back. Photo: Courtesy, The Science Museum, London.

Sunday, 25 December 1870 to coincide with the birthday of the ruling prince Rāmasiṃha. On the back of this astrolabe was engraved the entire fifth chapter of Mahendra Sūri's *Yantrarāja* (see Fig. 12.3). This mammoth piece could certainly not be used for observations. It was meant purely as a teaching aid to supplement the study of the text. But this instrument also incorporates the operational part of the text itself. Thus we have made a full circle by starting from Mahendra's text on the astrolabe and coming back to the same text this time engraved on an astrolabe. Sanskrit astrolabes continued to be produced in Rajasthan even up to 1902. A specimen made in this year is in a private collection at Paris.

2.7 The astrolabe thus played an important role in the history of science in medieval India—both as an observational and computational device, and as a graphic teaching tool—and testifies to the Indian astronomer's receptivity to valuable concepts from the outside. A full account of the history of this instrument in India together with descriptions of all extant specimens produced in India will appear in the final report of the present project.

ACKNOWLEDGEMENTS

This paper forms part of the ongoing project "A Descriptive Catalogue of Indian Astronomical and Time-Measuring Instruments," which is funded by the Indian National Science Academy and sponsored by the Indira Gandhi National Centre for the Arts. Grateful thanks are due to the authorities of these organisations.

APPENDIX

SANSKRIT TEXTS ON THE ASTROLABE

1. Mahendra Sūri, *Yantrarāja,* AD 1370.
2. Padmanābha, *Yantrakiraṇāvalī,* 1423.
3. Rāmacandra Vājapeyin, *Yantraprakāśa,* 1428.
4. Parameśvara, *Siddhāntadīpikā* on the *Mahābhāskarīya,* 1432.
5. Megharatna Muni, *Usturalāvayantra*, end of 15th c.
6. Visrāma, *Yantraśiromaṇi*, 1615.
7. Nṛsiṃha Daivajña, *Vāsanāvārttika* on the *Siddhānta-śiromaṇi,* 1621.
8. Nityānanda, *Sarvasiddhāntarāja*, 1639.
9. Anon, *Yantraprakāra,* ca. 1724-1740.
10. Jayasiṃha, *Yantrarājaracanā,* ca. 1724-1740.
11. Anon, *Sarvadesīya-Jarakālī-yantra,* ca. 1724-1740.
12. Anon, *Yantrarājavicāraviṃśādhyāyī,* ca. 1724-1740.
13. Śrīnātha Chagāṇī, *Yantraprabhā*, after 1740.
14. Nandarāma, *Yantrasāra,* 1772.
15. Mathurānātha Śukla, *Yantrarājaghaṭanā* or °*kalpa*, 1782.
16. Bāpudevaśāstrin, *Yantrarājopayogi-chedyaka*, mid 19th c.

13

Kaṭapayādi Notation on a Sanskrit Astrolabe

1.1 The *Kaṭapayādi* system is an ancient method of alphabetical notation where each consonant of the Sanskrit alphabet is given a numerical value.[1] The system is described in an anonymous line thus: *kādi nava, ṭādi nava, pādi pañca, yady aṣṭau,* "the nine [consonants] starting with *ka,* the nine starting with *ṭa,* the five starting with *pa* and the eight starting with *ya* [successively denote the numbers 1 to 9]." But the line does not say how the zero is to be represented. The *Sadratnamālā*, composed by Śaṅkara Varman in AD 1819, gives a comprehensive definition: "*na, ña* and the vowels are zero. The letters (of the consonants groups) commencing with *ka, ṭa, pa* and *ya* are digits. In conjunct letters the last consonant is to be taken as the digit. A consonant not attached to a

First published in: *Indian Journal of History of Science*, 34 (1999) 273-87.

1. The important literature on the *Kaṭapayādi* system is the following, arranged in chronological order. Whish, C.M., "On the Alphabetical Notation of the Hindus," *Transactions of the Literary Society of Madras,* 1827, 1, 54-62; Ojha, Gaurishankar Hirachand, *Bhāraṭīya Prācīna Lipimālā: The Palaeography of India,* [Delhi 1894; revised and enlarged second edition 1918]; reprint: Delhi 1971, p. 123; Fleet, J.F., "The Katapayadi System of Expressing Numbers," *Journal of the Royal Asiatic Society,* 1911, pp. 788-794; Datta, Bibhutibhusan & Singh, Avadhesh Narayan, *History of Hindu Mathematics: A Source Book,* [1935-38], 2nd edn, Bombay 1962, part 1, pp. 69-72; Raja, K. Kunjunni, "Astronomy and Mathematics in Kerala (An Account of the Literature)," *Adyar Library Bulletin,* 1963, 27, 118-167; Sarma, K.V., *A History of the Kerala School of Hindu Astronomy (In Perspective),* Hoshiarpur 1972, pp. 6-8; idem, "Word and Alphabetic Numerical Systems in India," A.K. Bag & S.R. Sarma (ed), *The Concept of Sunya,* New Delhi 2003, pp. 37-71.

vowel is to be ignored."[2] This may be graphically shown in the following table:

1	2	3	4	5	6	7	8	9	0
ka	*kha*	*ga*	*gha*	*ṅa*	*ca*	*cha*	*ja*	*jha*	*ña*
ṭa	*ṭha*	*ḍa*	*ḍha*	*ṇa*	*ta*	*tha*	*da*	*dha*	*na*
pa	*pha*	*ba*	*bha*	*ma*					
ya	*ra*	*la*	*va*	*śa*	*ṣa*	*sa*	*ha*		

Though neither of the definitions expressly states, the numerals represented in this system are read from the right to the left.[3] This is a neat and elegant method of expressing long numbers, more so because the chronograms, apart from the numerical value they represent, are otherwise also meaningful. An oft quoted example is of Narayaṇa Bhaṭṭa, a great Sanskrit poet of Kerala, closing his devotional poem *Nārayaṇīyam* with the expression *āyur-ārogyasaukhyam*. On the one hand it is a prayer for longevity (*āyur*), health (*ārogya*) and happiness (*saukhyam*); on the other it is a chronogram indicating the date of completion of the work, viz., 17,12,211 civil days from the beginning of the present Kali era.[4]

1.2 What is the antiquity of this system and the geographical extent of its use? Perhaps the earliest occurrence of this notation is in the *Candra-Vākyas* of Vararuci who is said to have lived in the fourth century AD.[5] In his commentary on the *Āryabhaṭīya,* Sūryadeva Yajvan persuasively argues that Āryabhaṭa must have

2. Cf. Sarma, K.V. "Word and Alphabetic Numerical System" (see n. 1).
3. Following the anonymous dictum *aṅkānāṃ vāmato gatiḥ,* "the numerical digits in a chronogram proceed towards the left." This is true also of the word notation, commonly known as the *Bhūtasaṃkhyā* system. However, nobody has so far tried to investigate how this "right to left" sequence of numerals in an otherwise "left to right" writing came into being.
4. Cf. Sarma, K.V., "Word and Alphabetic Numerical Sysems," (see n. 1).
5. Ibid. However, according to Pingree, David, *Jyotiḥśāstra: Astral and Mathematical Literature,* Wiesbaden 1981, p. 47, "the earliest attested epoch of the lunar *vākyas* is 1184."

known the *Kaṭapayādi* system,[6] thereby implying that the system was already prevalent in the fifth century AD. However, the first positive and datable occurrence is its use by Haridatta in his *Grahacāranibandhana,* composed in AD 683.

1.3 Regarding the geographical extent of its use, Kunjunni Raja observes: "The *Kaṭapayādi* system is well known only in South India and is most popular in Kerala. [...] It is generally believed to be one of the major contributions of Kerala to Indian mathematics."[7] That it was employed in Kerala very widely, not only in works on astronomy and mathematics, but also in non-scientific works, not only in Sanskrit writings but in Malayalam as well, is now quite well established.[8] Indeed, in order to facilitate its use in Malayalam texts, the purely Dravidian consonant *ḷ* was also incorporated into the system with a numerical value of 9.[9] Even in non-scientific works, the authors preferred to give the date of composition in terms of *ahargaṇa* in the *Kaṭapayādi* system. Since the number expressing the *ahargaṇa* is rather large, the *Kaṭapayādi* notation is more convenient than the word numerals; the added advantage being that the chronogram can be so formed as to yield some significant or charming meaning besides the numerical data. While the use of this system in Kerala has been well documented, very little information is available about other parts of South India; still less about the use of the system north of the Vindhyas.

1.4 In these circumstances, the discovery of a Sanskrit astrolabe[10] produced in north India on which the degrees of altitude are marked in the *Kaṭapayādi* notation is a significant datum in the history of this system. Even otherwise, this is an important specimen for the history of Indian astronomical instruments. While the majority of extant Sanskrit astrolabes contain a single plate

6. Cf. Sarma, K.V., (ed.), *Āryabhaṭīya of Āryabhaṭa, with the Commentary of Sūryadeva Yajvan,* New Delhi 1976, p. 10.
7. Raja, op. cit., p. 122.
8. Cf. Sarma, K.V., "Word and Alphabetic Numerical Systems in India."
9. Cf. Sarma, K.V., *A History of the Kerala School of Hindu Astronomy,* pp. 6-8.
10. On Sanskrit astrolabes in general, see Sarma, Sreeramula Rajeswara, "Yantrarāja: The Astrolabe in Sanskrit". *Indian Journal of History of Science,* 1999, 34, pp. 145-158; reprinted in this Volume, pp. 240-56.

calibrated for the use at a single terrestrial latitude, the present astrolabe is one of the few that contain multiple plates or tympans and it shows great affinity to the Mughal astrolabes of the sixteenth and seventeenth centuries. Therefore, this astrolabe deserves an independent treatment. In the following pages, a full technical description of the instrument will be given first and then the use of the *Kaṭapayādi* notation on it will be explained.

2.0 In connection with my project on "A Descriptive Catalogue of Indian Astronomical and Time-Measuring Instruments,"[11] I have studied the instruments preserved in the Sarasvati Bhavan Library of the Sampurnanand Sanskrit University, Varanasi. This library has one of the richest collections of Sanskrit manuscripts. It also possesses three astrolabes:

(i) Indo-Persian astrolabe with four plates, produced in the eleventh regnal year of Jahāngīr in AD 1616.[12]
(ii) Sanskrit astrolabe with five plates, anonymous, not dated but attributable to the seventeenth century—the subject of the present paper.
(iii) Sanskrit astrolabe with a single plate, anonymous and undated (but probably manufactured in the eighteenth century).[13]

There are also some interesting European instruments, such as a Gunter's Quadrant and a "synchronom" pendulum clock designed to measure not only hours and minutes but also *ghaṭīs* and *palas*.

2.1 The present astrolabe is made of brass. One of the sighting vanes attached to the alidade is broken and lost. Plate no. 2 has a crack running across two-thirds of the width parallel to the diameter.

11. On this project, see Sarma, Sreeramula Rajeswara, "Indian Astronomical and Time-Measuring Instruments: A Catalogue in Preparation," *Indian Journal of History of Science*, 1994, 29, pp. 507-528; reprinted in this Volume, pp. 19-46.
12. It was described, together with a photogaph, in Dube, Padmakara, "A seventeenth Century Astrolabe," in: Gopinath Kaviraj (ed.), *The Princess of Wales Sarasvati Bhavan Series,* vol. II, Benares 1923, pp. 129-136. Another photograph appeared in Bhaṭṭācārya, Vibhūti-bhūṣaṇa (ed.), *Yantrarāja-Vicāra-Viṃsādhyāyī*, Varanasi 1979.
13. See ibid., for a photograph.

Otherwise, the astrolabe is in a good state of preservation. It is of medium size with a diameter of 202 mm and a thickness of 9 mm (Fig. 13.1). The circular body or mater is surmounted by a triangular suspension piece known as *kursī* or "throne". The *kursī* is high, rising above the mater by 47 mm. It is lobed and scalloped, and culminates in a trifoliate top. To this top is attached a suspension mechanism consisting of a shackle and a ring. The trifoliate design is repeated in the terminals of the shackle as well. The *kursī* is cast in one piece with the mater and the upraised rim called limb. The front and back of the *kursī* are plain without any surface decoration. In Mughal astrolabes, the *kursī* is generally pierced, but there are a number of cases where it is solid but lobed and scalloped like this one.[14]

On the front side of the *kursī* is engraved the name *Paṃ. Rāmayatna Ojhā* in Devanāgarī script. This engraving is somewhat broader in comparison to the other engraved forms of writing on the astrolabe. It is probably a later addition, indicating the name of a subsequent owner of this astrolabe and not necessarily of the one who got it originally manufactured for his use.[15] Leaving this aside, there are yet three distinct styles of engraving of letters and numerals, to be found respectively in (i) the geographical gazetteer, (ii) altitudes written in the *Kaṭapayadī* notation on the plates, and (iii) the rest. The first two styles are somewhat akin to one another with a common characteristic form of the letter *ra.*

The limb of the mater contains a double band of scales. The inner band is graduated for each degree, while groups of 5° are marked on the outer band; these are labelled in Devanāgarī numerals as 5, 10, 15.....90, separately for each quarter, starting from the east or west point and proceeding to the south or north point.

14. For example, an astrolabe made by Mullā 'Īsá ibn Allāhdād (*fl.* AD 1601-16), now in a private collection in UK. It is illustrated in Wynter, Harriet & Anthony Turner, *Scientific Instruments*, London 1975, pp. 16-18. Astrolabe no (i) of the present collection also has a solid *kursī.*
15. Upādhyāya, Baladeva, *Kāśī kī Pāṇḍitya Paramparā*, second edition, Varanasi 1994, pp. 307, 907, 909 mentions a Rāmayatna Ojha, who was the first Head of the department of Jyotiṣa in Banaras Hindu University. He died in 1938. But he is too recent to have been the owner of this astrolabe.

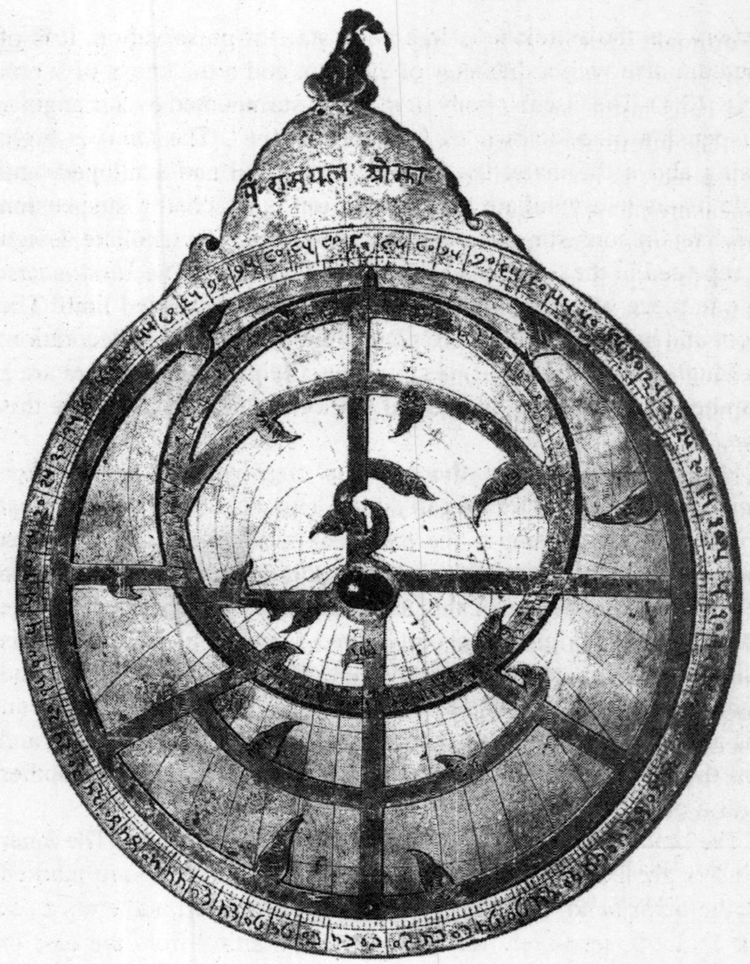

Fig. 13.1. Sanskrit University astrolabe; front view with the star map; below the perforated star map can be seen the plate with ecliptic coordinates.

2.2 Within the mater, the inner surface is divided into six annular rings. The first three are divided into 48 cells, while the three inner rings are divided into 24 cells each. The 48 cells of the outer circle are filled respectively with the names, longitudes and latitudes of 47 towns of the Indian subcontinent, Afghanistan and Iran; the 48th cell contains the argument. The longitudes are measured from the

Fortunate Isles (Arabic: *al-Jazā'ir al-Khālidāt*), roughly 35° west of Greenwich. Data for further 24 towns could have been filled in the 24 cells in the inner circle also, but these are left blank.

The information about the 47 towns, or the so-called geographical gazetteer, is not obtained by the astrolabe maker through his own measurements. It is often derived from earlier astrolabes or from the Islamic astronomical tables known as *Zījs*. Therefore, such gazetteers can indicate the path of transmission of this geographical knowledge. In the present case, it is very clear that the coordinates are derived from the Mughal astrolabes of the sixteenth and seventeenth centuries. In their catalogue of the astrolabes at the Smithsonian, Sharon Gibbs and George Saliba give a consolidated list of place names and their geographical parameters.[16] A comparison of the gazetteer of our Sanskrit astrolabe with this consolidated list shows a great degree of correspondence.

The following table reproduces the gazetteer, reading from the top and proceeding in the clockwise direction. Those marked with an asterisk have coordinates identical with those in the consolidated list of Gibbs-Saliba. If the consolidated list has a different value, it is noted in perenthesis. It will be noticed that most of the deviations occur in the case of longitudes and that several of these are just scribal errors. The place names marked with a plus sign do not occur in the Gibbs-Saliba list. These are verified from other sources.[17]

S.No	Place Name *nagaranāma*	Longitude *vistāra*	Latitude *akṣāṃśa*	Identification
1	Vagdāda	80;30 (80;00)	33;25	Baghdad
2	* Sīrāja	88	29;36	Shiraz
3	* Valaka	101	36;41	Balkh
4	Hareu	94;13 (94;20)	34;30	Herat
5	* Hosama	85;10	37	Hausam/ Khuzem

16. Gibbs, Sharon & Saliba, George, *Planispheric Astrolabes from the National Museum of American Hsitory*, City of Washington 1984, pp. 192-200.
17. Kennedy, E.S. & M.H., *Geographical Coordinates of Locations from Islamic Sources,* Frankfurt 1987; Habib, Irfan, *An Atlas of the Mughal Empire*, 2nd edn, Delhi 1986.

6	+	Katamī	85	36;55	Kutom
7		Kamdhāra	106;40 (107,40)	33	Qandahar
8		Mulatāna	106;35 (107;35)	29;40	Multan
9		Thaho	102;7	25;10	?
10		Huramuja	95;25 (92;00)	25	Hurmuz
11	*	Vadakasāna	104;24	37;10	Badakhshan
12		Kāsagara	106;3 (106;30)	44 (45)	Kashghar
13		Samarakamda	99;15 (99;16)	39;37	Samarqand
14	*	Vukhāro	97;30	39;50	Bukhara
15		Bhorura	108;40	31;10	?
16		Parasanūra	115;55 (85;55)	31;4	Peshawar
17		Ayodhyā	118;6 (108;6)	37;22 (27;22)	Ayodhya
18	+	Mānikapūra	118;10	26;49	Manikpur
19	*	Jaunapura	119;6	26;36	Jaunpur
20	*	Gopāmāū	116;33	26;45	Gopamau
21	*	Daulatāvādu	111	20;30	Daulatabad
22		Śamarakotu	105	25	?
23		Kāvila	104;40	34;30 (34;26)	Kabul
24		Sevāta	117;10	32;50	?
25		Ujjayinī	102 (110;5)	22;30	Ujjain
26	*	Vijayapura	105;3	17;20	Bijapur
27	*	Bhī(?)harāica	109;4 (111;05)	22;20 (21;20)	Broach
28	+	Caṃpānairi	108;45	22;30	Champaner
29	*	Ahamadābāda	108;40	23;15	Ahmedabad
30	+	Karo	117;6	26;35	Korah
31		Saṃbhara	115 (115;20)	28;6 (28;18)	Sambhal
32	*	Vadāū	114;59	27;32	Badaun
33		Āgarā	114	37;13 (27;13)	Agra
34	*	Thanesvara	112;33	3[0];10	Thanesar
35	*	Pānīpathā	113;20	28;52	Panipat
36	*	Kola	114;19	28;4	Aligarh
37		Syālakoṭa	109;4 (109;00)	33	Siyalkot
38	+	Varana	114	28;48	Baran
39	*	Suratānakoṭa	115	28;30	Sultankot
40		Dillī	113;35	28;29 (28;39)	Delhi
41		Ajameri	111;05	24 (26)	Ajmer
42	*	Gvāliyara	114	26;29	Gwalior
43	*	Kaśmīra	108	35	Kashmir
44	*	Tripapa	110	40	Tibet
45		Vārāṇasi	117;20 (107;20)	[2]6;55	Benares
46		Lāhaura	119;20 (89;20)	31;50	Lahore
47		Kanauja	115;50	26;35	Kannauj

2.3 The back of the astrolabe is divided into four quadrants by the vertical and horizontal diameters (see Fig. 13.2). The edge of the upper two quadrants is calibrated as in the front; i.e. the inner ring is divided into 90 degrees each and the outer ring into 18 parts which are labelled as 5, 10, 15 . . . 90, starting from the east and the west points and proceeding to the top. The edge of the two lower quadrants contains the cotangent scales as projected from the shadow squares. The upper left quadrant has engraved on it 60 parallel horizontal lines, each fifth being highlighted by a dotted line, and thus forms the sine graph. From the centre are drawn 18 radian lines for each 5°.

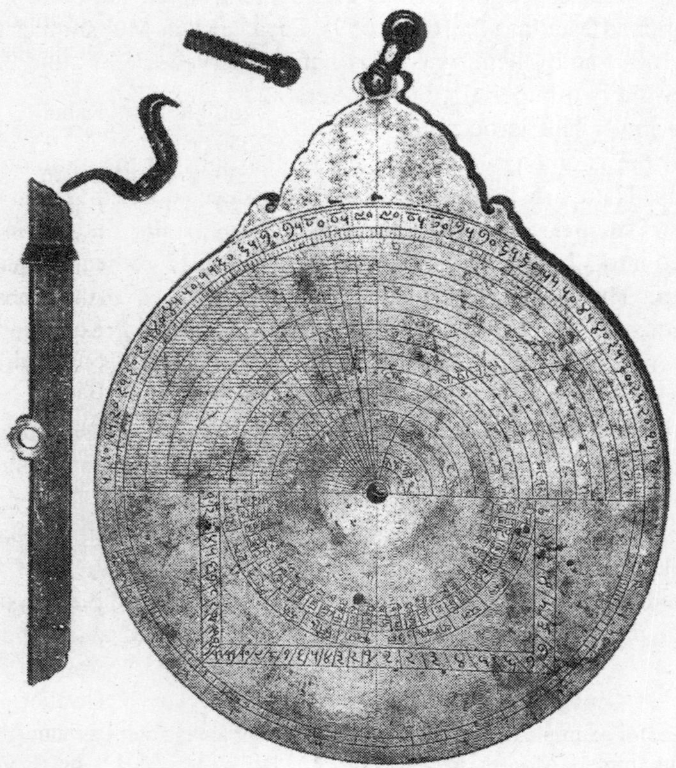

Fig. 13.2. Sanskrit University astrolabe; back view; with the alidade, nut and pin.

In the upper right quadrant are drawn 17 equidistant and concentric quarter circles. The space between two successive circles represents 10° of the zodiac. These spaces are labelled along the vertical and horizontal radii with the serial number and the first letter of the name of the zodiac sign in the following manner. On the vertical radius, starting from the top, 9 *dha*[*nuṣ*], 8 *Vṛ*[*ścika*], 7 *Tu*[*lā*], 6 *Ka*[*nyā*], 5 *Siṃ*[*ha*], 4 *Ka*[*rka*]; then on the horizontal radius, starting from the centre: 3 *Mi*[*thuna*], 2 *Vṛ*[*ṣabha*], 1 *Me*[*ṣa*], 12 *Mī*[*na*], 11 *Kuṃ*[*bha*], 10 *Ma*[*kara*].

Upon these arcs is projected the curve of the meridian altitude of the sun through the year; below the curve is written *Lāhaura* 32, indicating that the solar meridian altitudes pertain to the city of Lahore situated roughly at 32° N. In the Mughal astrolabes by Muḥammad Muqīm (*fl.* 1609-1659), Ḍiyā' al-Dīn Muḥammad (*fl.* 1645-1680) and others, one finds similar curves for 27° and 32°; i.e. for the two imperial cities of Agra and Lahore.[18]

The lower half is occupied by a double shadow square; on the left for the gnomon of 12 digits and on the right for the gnomon of 7 digits. There are, however, no labels on these squares. Within the shadow squares are two semi-circular bands; the outer one is divided into 12 parts in which the names of the 12 zodiac signs are written. The inner band contains the 28 names of the lunar mansions. Thus these two bands display the mutual correspondence between the 12 signs and 28 lunar mansions. This is also a characteristic feature of the Mughal astrolabes.

At the centre of the back is pivoted an ornate alidade (*vedhapaṭṭī*). It is 197 mm long and has a bevelled edge. Arcs corresponding to those in the upper right quadrant are engraved over half the length of the alidade. One of the two sighting vanes attached to the alidade is broken. The alidade together with the plates and the rete are attached to the mater by means of a large broad-headed pin passing through the central hole. A bird-shaped nut is screwed to the end of

18. See, for example, Fig. 2, showing the back of an astrolabe produced by Muḥammed Muqīm in the year AH 1031 (AD 1621), in Sarma, Sreeramula Rajeswara, "The Lahore Family of Astrolabists and their Ouvrage", *Studies in History of Medicine and Science,* 1994, 13, pp. 205-224.

the pin so that the alidade and the plates are held rightly in position.[19]

2.4 In the recessed space on the front side of the mater are a series of plates for various latitudes and on the top of them the star map called rete (Arabic *'ankabūt;* Sanskrit *bhapatra*) with a diameter of 181 mm (see Fig. 13.1). This is a circular open work plate, from which large portions have been cut off, leaving out the outer periphery constituted by the Tropic of Capricorn, the ecliptic circle and parts of the equatorial circle. These are held together by an east-west bar with two counter changes and a vertical bar with a single counter change.

The ecliptic circle is divided into 12 signs of the zodiac and labelled with the respective names in Sanskrit. Each sign is further subdivided into five parts of 6° each and labelled as 1, 2, 3, 4, 5. There are 19 star pointers, shaped like tiger's claws or stepped up tiger's claws. These arise from the circles and the horizontal and vertical bars. All but one of these are named. There is a small knob at the north point, with which the rete can be rotated around the centre.

The Mughal astrolabes contain highly ornate retes with floral traceries joining various star pointers. In comparison, the rete of the present astrolabe is rather simple and austere. But so is the rete in the very first extant Indo-Persian astrolabe, made by Allāhdād in AD 1567,[20] and the similarity between these two retes in respect of design and in respect of the fewer number of star pointers is quite striking.

Given below are the names of the stars marked in the rete of our astrolabe and their identification. The first seven are outside the ecliptic and the rest are within.

2.5 Nested within the hollow space of the mater are five circular plates or tympans (Arabic: *ṣafīḥa*; Sanskrit: *akṣapatra*) with projections engraved on both sides. One plate, slightly thicker than the others, carries the projections of the ecliptic co-ordinates on the

19. This is not the traditional practice in astrolabes. Usually the other end of the pin has a hole, into which a horse-shaped wedge is inserted.
20. It is with the Salar Jung Museum, Hyderabad; cf. Sarma, Sreeramula Rajeswara, *Astronomical Instruments in the Salar Jung Museum*, Hyderabad 1996, esp. pp. 7-10, Pls. 1-3 8, 11.

S.No.	Name of the Star in Sanskrit	Identification
1	Samudrapakṣī	ι Ceti
2	Rohiṇī	α Tauri
3	Dakṣiṇapāda	κ Orionis
4	Ārdrā-(?)	α Canis Maioris
5	Lubdhakabandhu	α Canis Minoris
6	Maghā	α Lionis
7	Citrā	α Virginis
8	Aśvinī	γ Arietis
9	Pretasara	β Persei
10	Skanda	α Aurigae
11	Āryamā	β Lionis
12	Svātī	β Boötis
13	Viśakhā Mātṛmaṇḍala	α Coronae Borealis
14	Dhanakoṭi (read: Dhanuḥkoṭi)	α Opiuchi
15	Abhijit	α Lyrae
16	Śravana	α Aquilae
17	Kukkuṭapakṣa	α Cygni
18	Pū[rva]bhā[drapadā]	β Pegasi

obverse and multiple horizons on the reverse. The other four plates serve eight different latitudes. The degrees of these latitudes (*akṣāṃśāḥ*) are engraved at the centre of the plate concerned. In three cases, the name of an important town situated on that latitude is also mentioned.[21]

1a	*akṣāṃśāḥ*	18	
1b	*akṣāṃśāḥ*	21	
2a	*akṣāṃśāḥ*	24	
2b	*akṣāṃśāḥ*	27	
3a	*akṣāṃśāḥ*	28	Vaṃśāvalīnagare
3b	*akṣāṃśāḥ*	25;39	(read 28;39) Yoginīpure (= Delhi)
4a	*akṣāṃśāḥ*	30	
4b	*akṣāṃśāḥ*	32	Lāhaura
5a	ecliptic co-ordinates		
5b	multiple horizons		

21. Together with the latitude, it is customary also to mention the maximum duration of the sunlight hours (*paramadina*) and sometimes also the length of equinoctial shadow (*chāyā*), but these are not mentioned here.

Of the first four plates, 1a, b; 2a, b; 3b; 4a, b are sexpartite, that is to say, on these plates equal altitude circles or almucantars (from the Arabic *al-muqanṭara*: Sanskrit: *unnatāṃśa-rekhā*) are drawn for each 6° (see Fig. 13.3). Plate 3a, however, is tripartite. Here equal altitude circles are drawn for each 3° (see Fig. 13.4). But in all the plates, azimuth lines are drawn for each 10°, that too only below the horizon. On the right hand side below the horizon is written *para* or *paścima* (west) and on the left *pūrva* (east). On all the plates, lines are drawn for seasonal or unequal hours, and these are numbered from 1 to 12, starting at the western horizon and proceeding towards the eastern.

3.1 *Kaṭapayādi* NOTATION ON THE ASTROLABE. It has been mentioned that on all the plates, with the exception of 3a, equal altitude circles are drawn for each 6°. Interestingly, these lines are numbered not in Devanāgarī numerals, but in the *Kaṭapayādi* notation. Thus here we have alphabetic notation for multiples of 6, from 6 to 90. Likewise plate 3a provides the *Kaṭapayādi* notation for multiples of 3 from 3 to 90. As in Mughal astrolabes, these arguments are marked on both the eastern and the western sides of the altitude circles, in such a manner that the numbers form an interesting pattern like a double arch. A consolidated list of the two sets of notations is given below. The few variants are shown in parenthesis.

3	*ga*	6	*ti*
9	*dhā*	12	*ropa* (*raya*)
15	*meke* (*mayā*)	18	*daya*
21	*kara*	24	*bhare* (*bhere, bhara*)
27	*sara*	30	*nāge* (*nāga, naga*)
33	*gala*	36	*cāge* (*cala*)
39	*dhala*	42	*rāghe* (*raghe*)
45	*mabha*	48	*dāghe* (*dabha*)
51	*kāme* (*kame*)	54	*bhāśe* (*ghāme*)
57	*sāśe*	60	*nata*
63	*gati*	66	*tata*
69	*dhata*	72	*rase* (*rasa*)
75	*masa*	78	*dasa*
81	*pādo*	84	*bhādo* (*bhehi*)
87	*sādo*	90	*nidhi*

It may be seen that a majority of forms terminates in the long

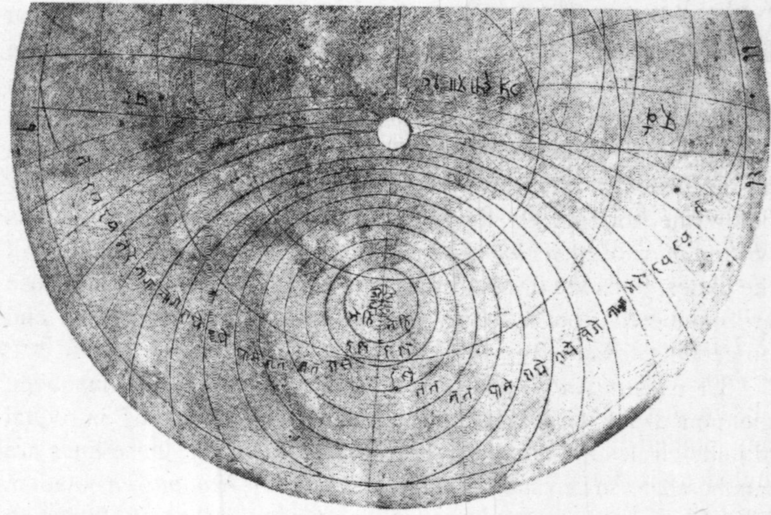

Fig. 13.3. Detail of the plate for latitude 18°, showing the altitude circles drawn for each 6° and labelled in the *Kaṭapayādi* notation.

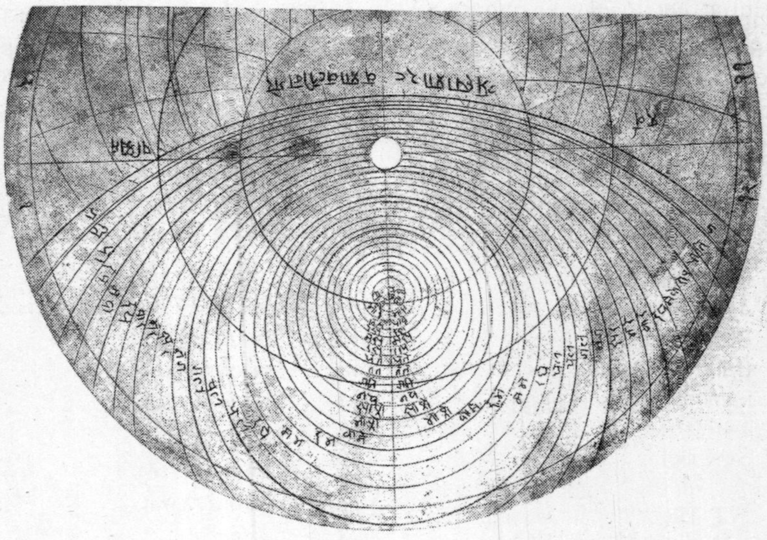

Fig. 13.4. Detail of the plate for latitude 28° Vaṃśāvalīnagara, showing the altitude circles drawn for each 3° and labelled in the *Kaṭapayādi* notation.

vowel *e*. It is probable that these were standard mnemonic forms to be recited in a sing-song fashion.

In Islamic astrolabes, whether produced in the west or east, all the numerical legends are written in the alphabetical notation known as *Abjad* notation.[22] In the course of this project, I have ascertained the existence of more than 70 Sanskrit astrolabes. Of these I have personally examined about 40 and collected full documentation for a further 16. Among these 56 and odd Sanskrit astrolabes, the present one is the only specimen to employ the *Kaṭapayādi* notation on the plates, in partial imitation of the Islamic astrolabes.

3.2 Besides the name of Pandit Rāmayatna Ojhā, the astrolabe does not carry any other inscription regarding its maker or the date of its manufacture. But, as has been mentioned above, there are many points of correspondence between this astrolabe and the Mughal astrolabes of the sixteenth and seventeenth centuries. Therefore, one is inclined to assign this astrolabe to the seventeenth century AD.

3.3 As noted above, in three plates the names of the towns have been mentioned, viz., Vaṃśāvalīnagara at 28° N; Yoginīpura, i.e. Delhi at 28:39° N and Lahore at 32°.[23] Thus the maker laid stress on these three towns. Of these, Delhi is, of course, the royal capital. We also saw that in the upper right quadrant on the back, there is a curve for the sun's meridian altitude for Lahore at 32°. It may suggest that the maker is from Lahore. On the other hand, while all other plates were sexpartite, that for Vaṃśāvalīnagara at 28° is tripartite, thereby allowing finer measurements of altitude at this

22. On *Abjad* notation, especially on the astrolabes, see Khareghat, M.P., *Astrolabes*, M.P. Khareghat Memorial Volume II, edited by Dinshaw D. Kapadia, Bombay 1950, pp. viii-xiii. In the last century, Ghulām Qādir of Kapurthala started using the Arabic numerals, i.e., the numerals connected with the Arabic/ Persian (and Urdu) script in his astrolabes. One astrolabe made in 1861 is in the National Museum, New Delhi. For an illustration, see Gupta, S.P. (ed), *Masterpieces from the National Museum Collection*, New Delhi 1985, p. 179, item no. 272. The National Museum has a few other astronomical instruments with similar numerals. Though the instruments are not signed, they may be attributed to Ghulām Qādir for this and other stylistic reasons.
23. The gazetteer mentions Dillī and not Yoginīpura and assigns to Lāhaura a more accurate latitude of 31;50°.

place. Can it then be that the maker is from this Vaṃśāvalīnagara? However, in spite of painstaking search, it has not been possible to locate any mention of this place situated, roughly on 28° N, either in Gujarat, or in Rājasthan, or in western Uttar Pradesh. It is possible that the place is now known under a Middle Indic form of the name that is phonetically far removed from Vaṃśāvalīnagara.

3.4 Be that as it may, this astrolabe demonstrates that the *Kaṭapayādi* system of representing numerals was not just confined to South India; it was known well enough in North India to be employed in an astrolabe. A fresh search in epigraphic records may yield further examples.

ACKNOWLEDGEMENTS

This paper forms part of the ongoing project "A Descriptive Catalogue of Indian Astronomical and Time-Measuring Instruments", which is funded by the Indian National Science Academy and sponsored by the Indira Gandhi National Centre for the Arts. Grateful thanks are due to the authorities of these organisations. I am also highly obliged to Professor Mandana Misra, the then Vice-Chancellor of Sampuranand Sanskrit University, Varanasi, for permission to study the astronomical instruments preserved in the Sarasvati Bhavan Library; Shri D.S. Mishra, Assistant Librarian, and his staff at the Sarasvati Bhavan Library for their warm-hearted cooperation; Professor R.C. Sharma, Director of the Bharat Kala Bhavan Museum of Banaras Hindu University, for various acts of kindness; and to Shri P.S. Prakash Rao, Bharat Kala Bhavan, for taking the photographs published with this paper.

PART IV

THE CELESTIAL GLOBE

14

From al-Kura to Bhagola: On the Dissemination of the Celestial Globe in India

1. Introduction

In the context of the transmission of astronomical instruments from the Islamic World to India between the 14th and 18th centuries, I wish to examine in this paper the dissemination of the celestial globe and its reception by Hindu astronomers.

Described quite elaborately by Ptolemy in his *Almagest* (VIII.3), the celestial globe is a convenient instrument for mapping the star positions and storing them.[1] It was adopted both in medieval Europe and in the Islamic World. In Europe, it came to be treated as the emblem par excellence of the astronomer just as the astrolabe was considered his chief symbol in the Islamic World. European portraits of the astronomer, like those of Copernicus or Kepler, usually show him plotting the star positions on the globe by means of a pair of compasses. During the Renaissance, automatically rotating globes with clockwork mechanism were made in large numbers and these represented the highpoint of technology as well as the goldsmith's craft.

In the Islamic World several treatises were written on this globe and innovations were made in its construction. In Arabic, it is called *al-Kura* "the sphere", *al-Bayḍa* "the egg" or *Dhāt al-Kursī* "the instrument resting in a horizontal frame".[2] In the early 12th century, al-Khāzinī designed a globe that rotated automatically by

First published in: *Studies in History of Medicine and Science*, 13.1 (1994) 69-85.

1. King, pp. 4, 18.
2. Ibid, p. 4.

means of a falling weight in a leaking reservoir of sand.[3]

The earliest extant globe in the Islamic World was manufactured by Ibraham ibn Sa'īd in Moorish Spain in 473 AH/AD 1080.[4] Mu'yyad al-Dīn al-'Urḍī, the celebrated instrument-maker of the Marāgha observatory, does not include it in his treatise on instruments,[5] but a specimen crafted by his son Muḥammad ibn Mu'ayyad al-'Urḍī in AD 1278 is preserved in Dresden.[6] Made by joining two brass hemispheres, it is engraved and inlaid with gold and silver. It can be made to rotate around the equatorial axis as well as the ecliptic axis. Apparently a similar model was taken from Marāgha to the court of the Mongol emperor Kublai Khān by a certain Jamāl al-Dīn in AD 1267 together with many other astronomical instruments.[7] Al-Saman describes what is supposed to be a second extant globe by Muḥammad ibn Mu'ayyad al-'Urḍī.

In India, automatic globes were envisaged as teaching instruments. Towards the end of the fifth century AD, Āryabhaṭa speaks of an automatically rotating (*svayaṃvaha*) globe (*gola*), the motive power for which is provided by a sinking float in an outflow type of water clock.[8] However, this globe was not intended to map star positions but only to demonstrate apparent motion of the great circles. The real celestial globe with star positions marked on it is a late-comer to India. It was introduced in this country by the Mughal emperor Humāyūn in the middle of the 16th century.

2. Humāyūn's Interest in Astronomy

Humāyūn (reign 1530-1556) was greatly interested in astronomical instruments.[9] Abū 'l-Fazl, minister and chief chronicler of his son Akbar, often speaks in glowing terms of Humāyūn's interest in

3. Lorch.
4. ESS, p. 214.
5. Seeman.
6. The present writer had the opportunity of seeing this remarkable globe at a special exhibition at Linz, Austria. It was described and illustrated in the catalogue of this exhibition, cf. Seipel, vol. II, p. 20.
7. Hartner, pp. 215, 221.
8. Sarma (a), pp. 69-71.
9. Nadvi (b), p. 269.

astronomy. Thus he says at one place that "His Majesty, who in astrolabic investigations and studies in astronomical tables and observations was at the head of the enthroned ones of acute knowledge and who was a second Alexander. . . ."[10] Elsewhere, employing an astronomical metaphor, he calls Humāyūn "the alidad of the astrolabe of theory and practice."[11]

When Humāyūn lost the throne of India and wandered through Persia and Iraq, his constant companions were astronomers. The chronicles never tire of declaring that he was quite adept in the use of the astrolabe. His sister reports that when he became impatient of waiting for Hamida Bano's consent to marry him, he took the "astrolabe into his blessed hands" and himself fixed the auspicious moment for their wedding.[12]

Planetary symbolism pervaded his daily routine, the decor of his court and even his amusements. He is said to have invented a kind of astrological monopoly game where, instead of squares, there were circles representing the nine cosmic spheres. These circles were painted in the respective colours attributed to the spheres. The courtiers acted as the pawns and moved from one circle to the other according to the throw of dice.

Astronomy, as is well known, was also the cause of Humāyūn's untimely death. Waiting one evening in his library for the appearance of Venus, he heard the call for prayers and rushed down the stairs. In his haste, he slipped and fell, and succumbed to his injuries.

His interest in the celestial globe is revealed in an anecdote narrated by Abū 'l-Fazl:[13] "When Humāyūn reached Tabrīz during his journey in Persia, he ordered his slave Pik Muḥammad Akhta Begi to search for a celestial globe—*kura*—in that city, as he was very keen about astrolabes, globes and other observatory instruments. In Persian *kura* means a colt, so his simple slave obeyed the orders by bringing a number of colts to the royal presence. The king laughed when he saw the multitude of colts and mares before him, and bought them all the same as a good omen."

10. H. Beveridge, p. 123.
11. Ibid, pp. 283-284.
12. A.S. Beveridge, p. 51.
13. H. Beveridge, p. 445.

3. Allāhdād the Royal Astrolabist

Under the active patronage of Humāyūn, astrolabes and celestial globes began to be manufactured in India. Astrolabes must have been made even before this time, though no actual specimens are extant today. The existence of several Sanskrit manuals on the construction and use of the astrolabe[14] testifies to the fact that this versatile instrument, called therefore in Sanskrit *yantrarāja* "king of astronomical instruments", must have been used both by Muslims and Hindus/Jains at least from the reign of Fīrūz Shāh Tughluq in the second half of the fourteenth century.

Fīrūz promoted, probably for the first time, manufacture of astrolabes in India. He also caused manuals to be written on the construction and use of the astrolabe both in Persian in Sanskrit.[15] The first Sanskrit manual on the astrolabe, composed at his instance in AD 1370 by the Jaina monk Mahendra Sūri is extant.[16] The Persian manual is no more extant but its summary survives in the anonymous chronicle *Sīrat-i Fīrūz Shāhi*, which was also composed in 1370. There is, however, no mention of the celestial globe in India before Humāyūn's time.

The *Ā'īn-i Akbarī* reports that the astrolabes and globes produced by Maulānā Maqṣūd of Herat, one of the servants of Humāyūn, were much admired by cognoscenti,[17] but none of them seems to have survived.

Humāyūn's chief astrolabist, however, was Master Allāhdād of Lahore, who signed his creations as Ustād Shaykh Allāhdād Āsṭurlābī Humāyūnī Lāhūrī. He and his descendants of the next four generations were the chief producers of astrolabes and celestial globes in India. Some 80 astrolabes and more than 20 celestial globes bear the signatures of the various members of this family. Besides these, a number of unsigned pieces are also attributed to this family on stylistic grounds. In 1935, Syed Sulaiman Nadvi[18] made for the first time a survey of the instruments produced by the

14. Sarma (b), pp. 238-241.
15. Sarma (e).
16. Raikva.
17. Blochmann, pp. 54-55.
18. Nadvi (a), pp. 621-631.

members of this family. More recently in 1985, Emilie Savage-Smith, in her exemplary study of the Islamic globes,[19] discussed in great detail those manufactured by the Allāhdād family.

Three astrolabes made by Allāhdād survive today. The only dated exemplar among these was produced in 975 AH/AD 1567, and is preserved today in the Salar Jang Museum of Hyderabad.[20] The other two, presumably made at about the same time, are now in the Billmeir Collection at the Museum of History of Science, Oxford.[21]

But we do not know of any celestial globe manufactured by the patriarch of the Lahore astrolabist family. The earliest celestial globe extant today was manufactured in 998 AH/AD 1589, during the reign of Akbar by one 'Alī Kashmīrī ibn Luqmān,[22] who is not a member of the Allāhdād family. Next in chronological order are the four globes created by Allāhdād's grandson Qā'im Muḥammad ibn Mullā 'Īsá between the years 1622 and 1637, i.e. during the reigns of Jahāngīr and Shāh Jahān. The first of these is at Stonyhurst College, Lancashire, UK. The earliest globe available in India is the one crafted by Qā'im in 1047 AH/AD 1637. It is now in the Khuda Bakhsh Oriental Public Library, Patna.[23] Qā'im's brother, Muḥammad Muqīm is known through 32 astrolabes[24] and one celestial globe.[25] Muqīm's son, Ḥamīd made at least one globe in 1065 AH/AD 1655, which is now in the Whipple Museum of History of Science, Cambridge, UK.[26]

4. Ḍiyā' al-Dīn Muḥammad

The most prolific and versatile member of this family is, however, Ḍiyā' al-Dīn Muḥammad son of Qā'im Muḥammad. Twenty-six astrolabes, produced by him between the years 1637 and 1680, are

19. Emilie Savage-Smith, *Islamicate Celestial Globes. Their History, Construction and Use,* Washington, D.C., 1985.
20. CCA 1120.
21. CCA 1089 & 2530; Anderson, p. 35, no. 126.
22. ESS 10.
23. Nadvi (a), p. 627; ESS 14.
24. CCA, s.v.
25. ESS 15.
26. ESS 68.

known today.[27] The last of these is not a conventional planispheric astrolabe but a very large universal astrolabe, invented originally by al-Zarqālluh of Cordova in the 11th century. Ḍiyā' al-Dīn's *Zarqālī* astrolabe, measuring 555 mm in diameter, is the only specimen of its kind in India. It is now preserved in Sawai Jai Singh's observatory at Jaipur.[28]

More particularly, Ḍiyā' al-Dīn Muḥammad produced the largest number of Islamic globes attributable to a single instrument maker. We know of 14 globes manufactured by him in the years between 1645 and 1680. These are now scattered all over the world: five in the United Kingdom, three in India, two each in Egypt and the USA, and one each in Russia and Germany. (See the table below).

Celestial Globes by Ḍiyā' al-Dīn Muḥammad

1	1055 AH/AD 1645	Columbia University, New York.
2	1057/1647	Asian Museum, St. Petersburg.
3	1058/1648	Private Collection, Patna.
4	1060/1650	Victoria & Albert Museum, London.
5	1064/1654	Aligarh Muslim University, Aligarh.
6	1067/1656	Victoria & Albert Museum, London.
7	1068/1657	Welsh Industrial & Maritime Museum, Cardiff.
8	1068/1657	Museum of Islamic Art, Cairo.
9	1070/1659	Museum of Islamic Art, Cairo.
10	1071/1660	Staatliche Museen, Berlin.
11	1074/1663	Royal Scottish Museum, Edinburgh.
12	1074/1663	Museum of the History of Science, Oxford.
13	1087/1676	Archaeological Museum, Red Fort, Delhi.
14	1090/1679	Time Museum, Rockford.

Again, the last of these is not a conventional celestial globe but an open-work sphere in which the spaces have been cut out, leaving the constellation figures and the great circles. The star positions are shown by holes filled with glass or mica. A light placed inside this globe would illumine the star points, constellation figures and the great circles. Thus this globe resembles the perforated brass lamp shades, the manufacture of which also began in Mughal times and still continues at Varanasi and Muradabad.[29]

27. CCA, s.v.
28. Kaye (a), pp. 27-30; see also Sarma (d).
29. ESS 35, Fig. 17.

The other thirteen conventional globes can be divided into two groups. Three of them show only the major or astrolabe stars. In no. 3, large inlaid silver points indicate some 20 star positions, while in nos. 11 and 13 engraved dots within small circles represent 60 and 92 stars respectively. The remaining globes, on the other hand, are highly complex ones, showing through inlaid silver points approximately 1018 star positions according to Ulūgh Beg's tables. Furthermore, the outlines of 48 constellation figures, mentioned by Ptolemy, are finely engraved.

While the common practice in the manufacture of globes is first to make two hollow hemispheres and then to join them, the globes of Qā'im Muḥammad and his son Ḍiyā' al-Dīn Muḥammad were cast in one single piece by *cire perdue* or lost wax process, which is more time consuming but artistically more challenging. During this process, large holes had to be kept open which were later closed with plugs of the correct size and the engraving was continued over them. Thus in most of these globes plugs can be seen near the north and south poles.

Savage-Smith describes almost all of Ḍiyā' al-Dīn's known celestial globes, save no. 5 which is at Aligarh. Therefore, this particular globe will be described more elaborately here.

5. The Aligarh Globe

While a number of Ḍiyā' al-Dīn's magnificent astrolabes are preserved in India,[30] only three of his globes are reported to be still in India. Of these three, the present whereabouts of no. 3 are not known; it was in private possession in 1935.[31] Globe no. 13 in the Archaeological Museum at Delhi is also a small piece showing only 92 stars.[32] Therefore, no. 5 is the only globe available in India which contains the full complement of star positions and constellation figures.[33]

30. Sarma (c).
31. Klueber.
32. Kaye (b), pp. 16-19.
33. Since writing this, I have located one more globe, dated AH 1074 (= AD 1663-64), in the Salar Jung Museum, Hyderabad; see Sreeramula Rajeswara Sarma, *Astronomical Instruments in the Salar Jung Museum*, Hyderabad, 1996, pp. 24-26, Pl. 13.

According to Nadvi,[34] this globe formerly belonged to a Ḥakīm of Rampur. Now it is in the library of the Ajmal Khan Tibbia College, Aligarh Muslim University (Fig. 14.1).

The globe measures about 122 mm in diameter and rests on a horizontal frame, where two mutually perpendicular rings must have been affixed vertically in order to represent the meridian circle and the prime vertical. These rings are now missing. The horizontal ring itself had broken into four pieces at the four indents and was later repaired by nailing copper strips at the broken places (Fig. 14.2).

In the globe there are two small holes at the two poles of the equator, through which the axis had passed originally, and was pivoted to the meridian ring. The axis is now missing and the globe is held above the frame by means of an iron wire. Close to each pole, there is a circular patch or plug. The one near the north pole measures 29 mm and that near the south pole is 40 mm in diameter. The latter can be seen in Fig. 14.3. The constellation figure of Centaurus extends over this plug. The equator and the ecliptic are marked by double bands of lines; one band is graduated in single degrees and the other in groups of five degrees and numbered (Fig. 14.4). Likewise the horizontal ring is also marked with a double band of lines. Single lines represent the tropics and the polar circles. Six great circles cut across the ecliptic perpendicularly and converge at its poles, thus dividing the globe into 12 segments or zodiac signs (Fig. 14.5).

About 1018 star positions are marked on the globe with inlaid silver points of three different sizes. The outlines of the 48 constellation figures and zodiac signs are meticulously engraved. Around the southern polar circle is the globe-maker's inscription (Fig. 14.5), which reads as follows:

"The work of the humblest creature Dīyā' al-Dīn Muḥammad, son of Qā'im Muḥammad, son of Mullā 'Īsá, son of Shaykh Allāhdād Āsṭurlābī Humāyūnī Lāhūrī, dated AH 1064."[35]

34. Nadvi (a), p. 628.
35. Ibid, p. 628.

Fig. 14.1. Aligarh globe with stand.

Fig. 14.2. Aligarh globe: horizontal frame.

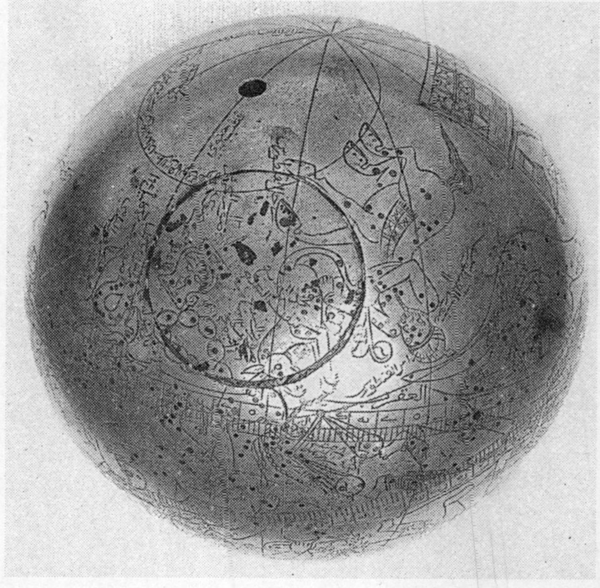

Fig. 14.3. Aligarh globe: plug near the South Pole.

Fig. 14.4. Aligarh globe: ecliptic and equator.

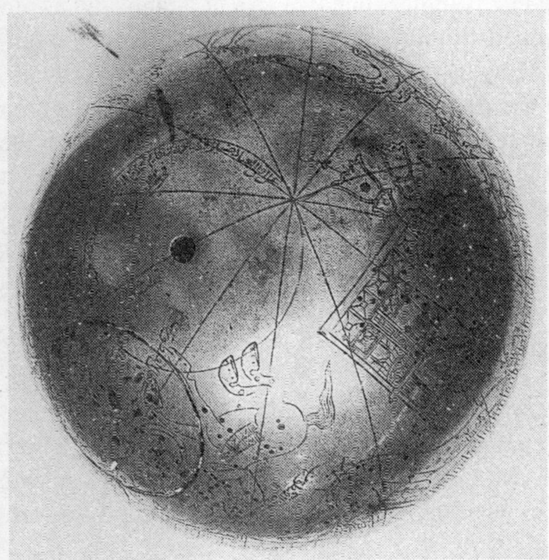

Fig. 14.5. Aligarh globe: maker's signature within the southern polar circle.

6. The Celestial Globe in Mughal Miniatures

The long period of Dīyā' al-Dīn Muḥammad's creative activity, viz., from AD 1637 to 1680, encompasses the reigns of Shāh Jahān (1628-1658) and Aurangzeb (1658-1707). In Shāh Jahān's time, the celestial globe had become popular enough to be depicted along with other astronomical instruments in two miniature paintings.[36] In both these miniatures, the astronomical instruments seem to transcend their physical function and acquire symbolic significance.

In the first miniature (Fig. 14.6), originally from an album made for Shāh Jahān and now in the Musée Guimet, Paris, a venerable scholar is portrayed amid idyllic nature. He is surrounded by his pupils and three astronomical instruments: a celestial globe to his right, a small astrolabe in front of him and a sand clock to his left. The whole ambience suggests that he is not just a professional astronomer but a saintly recluse living far away from the bustle of mundane life and that the instruments are not mere measuring tools but emblems of secret knowledge which reveal the cosmic mysteries to this astronomer-philosopher.

The second miniature (Fig. 14.7) is from a series of portraits of his forefathers commissioned by Shāh Jahān and is now at the Smithsonian Institution, Washington. On the top margin of each portrait, two angels are depicted holding various kinds of symbolic objects. In the portrait of Humāyūn with which we are concerned here, two angels are holding a crown over Humāyūn, suggesting his universal sovereignty. Besides the crown, one angel is holding a globe, which can only be a celestial globe, and the other is holding a ring dial, both obviously as symbols of cosmic Space and Time.

7. Other Globe-Makers

Outside the family of Allāhdād, we know of only one contemporary astrolabist cum globe-maker. This was Muḥammad Ṣāliḥ of Thatta, known from two astrolabes.[37] Till recently, only two celestial globes made by him were known. The first one was produced in 1070 AH/

36. Sarma (b), pp. 247-248; Pls. 6, 9.
37. CCA 23, 2502.

Fig. 14.6. Astronomer-philosopher surrounded by astronomical instruments. Detail from a Mughal miniature. See also Fig. 4.9.

Fig. 14.7. Angels holding the crown, celestial globe and ring dial. Detail from a portrait of Humāyūn.

AD 1659-60 for Shaykh 'Abd al-Khāliq.[38] In 1928 it was acquired by the Archaeological Museum at the Red Fort of Delhi, where it is now preserved. It measures 210 mm in diameter and is mounted horizontally on a stand which is clearly a late addition. It shows the full set of 1018 stars through inlaid silver points and the 48 constellation figures. There are holes at the equatorial and the ecliptic poles so that the globe can be made to rotate around the equatorial or the ecliptic axis.

The other globe by Muḥammad Ṣālih was crafted in 1074 AH/AD 1663 and is now in a private collection at London.[39] A remarkable feature in this globe is that the numbers on the scales, the names of the zodiac signs and other constellations and the maker's inscription are given in both Arabic and Sanskrit (in Devanāgarī script). Two more globes, one signed by Ṣāliḥ and another unsigned but attributable to him, were noticed by the present author and his colleagues and reported elsewhere.[40]

Ṣālīḥ's globes were also cast by the *cire perdue* process, which was initiated by Qā'im Muḥammad and his son Ḍiyā' al-Dīn Muḥammad. This tradition apparently continued up to the middle of the 19th century, though there are no extant globes cast in the intervening period. The last example of this tradition is a globe manufactured by Lālah Bahlūmal Lāhūrī in 1842 for the ruler of Kapurthala.[41]

Aside from Bahlūmal, two others distinguished themselves in the 19th century with their keen interest in Islamic astronomical instrumentation. Nawāb Zayn al-'Ābidin of Delhi was a mathematician and instrument maker. His nephew Sir Syed Ahmad Khan writes that Zayn al-'Ābidin constructed beautiful globes, astrolabes, armillary spheres and quadrants and that his studio containing these instruments looked like an observatory.[42] Unfortunately, none of these instruments are extant.

Of course, in the 19th century, Muslim astronomers of India began to be acquainted with European astronomy and astronomical

38. Dhama.
39. ESS 29, Fig. 18.
40. Sarma, Ansari & Kulkarni, pp. 80-88.
41. ESS 33, Fig. 24, pp. 52 ff.
42. Nadvi (b), p. 269.

instruments. Such awareness can be seen in Ghulām Ḥusain Jaunpūrī's encyclopaedia *Jāmi'-i Bahādur Khānī*, published from Calcutta in 1835. But the illustration of the celestial globe printed in this book[43] shows a traditional model. Indeed, a beautiful celestial globe manufactured by him in 1816 is in a private collection at Aligarh.[44]

These in brief are the Islamic celestial globes produced in India. There ought to be scores more, considering that the celestial globe, like the astrolabe, had been an important teaching aid for astronomy in the traditional *madrasahs*. Therefore, a thorough search needs to be undertaken, especially among Indian holdings.[45]

8. Hindu Astronomers and the Celestial Globe

What was the impact of this celestial globe on Hindu astronomers? It is, of course, not comparable to the astrolabe in its utility. Even so, it was noticed by Hindu astronomers even before Ḍiyā' al-Dīn's time. Nṛsiṃha Daivajña, hailing from a family of distinguished astronomers of Varanasi, wrote a commentary on Bhāskara's *Siddhāntaśiromaṇi* in 1621. In this commentary he gives a detailed description of the celestial globe which he calls *bhagola*.[46] He explains that the Muslims obtain the star coordinates by direct observation and then mark these positions on the globe. He goes on to say that the Muslims call this instrument *kura*, and that the horizontal frame, which resembles the stand used by the Hindus for placing their sacred conch shell, is known as *kursī*.

The Hindu astronomers, however, did not make much use of this instrument. Consequently, there are very few extant specimens with legends written in Sanskrit language and Devanāgarī script. I have seen so far only three. The first of these may have been made for Sawai Jai Singh and is now at the Hawa Mahal Museum, Jaipur

43. Jaunpuri, p. 515.
44. Ansari & S.R. Sarma.
45. The present writer commenced such a survey in 1991 and has catalogued the instruments in most of the collections in the US and UK, and also of several museums in India, the reports of which are in preparation and will be published in the coming years.
46. Caturveda, p. 438.

(Fig. 14.8). Also at Jaipur, the Museum of Indology possesses a copper globe, on which just the ecliptic and the names of the zodiac signs in Sanskrit are marked. The third globe once belonged to David Eugene Smith,[47] the well known historian of mathematics, and is now in the Butler Library, Columbia University , New York. It is not signed nor dated, but Savage-Smith attributes it to the

Fig. 14.8. Sanskrit celestial globe from Jaipur.

47. For a drawing of this globe, see Smith, vol. II, p. 365, where it is assigned to ca. 1600.

workshop of Lālah Bahlūmal Lāhūrī of mid-nineteenth century.[48]

A vigorous search may yield a few more Sanskrit globes, but on the whole their number is nowhere near that of the Arabic/Persian globes produced in India, or for that matter, nowhere near the number of Sanskrit astrolabes. The reason is quite obvious. Hindu astronomers were interested in the coordinates of a very limited number of stars and not in all the 1018 stars marked on the celestial globe. Nṛsiṃha Daivajña says that only the lunar mansions and a few other constellations like Canopus should be marked on the astrolabe. "Only these can be recognised by Hindu scholars. The stars known to the Muslims do not serve our purpose. Observation of unfamiliar stars would only lead to misfortune." [49]

The one person who tried to bridge this cultural gap was Sawai Jai Singh of Jaipur. In the 1730's he ordered a Sanskrit translation of Naṣīr al-Dīn al-Ṭūsī's Arabic version of the *Almagest,* including the description of the celestial globe,[50] a description with which in fact the history of this instrument commenced. Thus we have come a full circle, perhaps the proper thing to do in matters concerning the globe.

ACKNOWLEDGEMENTS

Grateful thanks are due to Hakim Qiyamuddin (Principal) and Professor Hakim S. Zillurrehman (Chairman, Department of Ilmul Advia) of the Azmal Khan Tibbia College, Aligarh Muslim University, for enabling me to study the "Aligarh Globe"; to Professor Owen Gingerich, Professor of Astronomy and History of Science, Harvard-Smithsonian Center for Astrophysics, Cambridge, Mass., for advice on Ḍiyā' al-Dīn's oeuvre and for the generous gift of literature not accessible to me.

48. ESS 54, Fig. 25.
49. Caturveda, p. 449.
50. Sharma, vol. II, pp. 781-783.

BIBLIOGRAPHY

Al-Saman, T.: "A Second Celestial Globe by Muḥammad ibn Mu'ayyad al-'Urḍī?" read at International Colloquium on Interaction of European and Asian Astronomy, Vienna, 1990.

Anderson, R.G.W.: *Science in India: A Festival of India Exhibition at the Science Museum, London, Catalogue*, London, 1982.

Ansari, S.M.R. & Sarma, S.R.: "Ghulām Ḥusain Jaunpūrī's Encyclopaedia of Mathematics and Astrononomy," *Studies in History of Medicine and Science*, 16 (1999-2000) 77-93.

Beveridge, A.S.: *The History of Humāyūn (Humāyūn-Nāma by Gul-Badan Begum)*, Delhi, 1972.

Beveridge, H. (tr): *Akbarnāma of Abūl Fazl*, vol. I, Calcutta, 1910.

Blochmann, H. (tr): *Ā'in-i Akbarī of Abūl Fazl,* vol. I, reprint: New Delhi, 1977.

Caturveda, Muralidhara (ed): *Siddhānta-Śiromani of Bhāskara, with the Vārttika by Nrsimha Daivajña,* Varanasi, 1981.

CCA—Gibbs, Sharon L., Henderson, Janice A. & Price, Derek de Sola: *A Computerized Checklist of Astrolabes,* Yale University, New Haven, 1973. The references are to the Check-list numbers.

Dhama, B.L.: "New Acquisitions in the Delhi Museum," *Annual Report of the Archaeological Survey of India*, 28 (1928-29) 143-144.

ESS—Savage-Smith, Emilie: *Islamicate Celestial Globes: Their History, Construction and Use*, Washington, D.C., 1985. References are to the catalogue numbers, fig(ures), or p(ages).

Hartner, Willy: "The Astronomical Instruments of Cha-ma-lu-ting, their Identification, and their Relations to the Observatory of Marāgha" in: idem, *Oriens-Occidens*, vol. I, Hildesheim, 1985.

Jaunpuri, Ghulam Husain: *Jāmi'-i Bahādur Khānī*, Calcutta, 1835.

Kaye, G.R. (a): *The Astronomical Observatories of Jai Singh*, Calcutta, 1918.

Kaye, G.R. (b): *Astronomical Instruments in the Delhi Museum,* Calcutta, 1921.

King, David A.: "Astronomical Instrumentation in the Medieval Near East" in: idem, *Islamic Astronomical Instruments*, London, 1987.

Klüber, Harald von: "Über einen Arabischen Himmelsglobus aus Indien," *Baesler Archiv: Beitraege zur Völkerkunde*, 18 (1935) 1-21.

Lorch, Richard: "Al-Khāzinī's Sphere that Rotates by Itself," *Journal for the History of Arabic Science*, 4 (1980) 287-329.

Nadvi, Syed Sulaiman (a): "Some Indian Astrolabe Makers," *Islamic Culture*, 9 (1935) 621-631.

Nadvi, Syed Sulaiman (b): "Muslim Observatories," *Islamic Culture*, 20 (1946) 267-281.

Raikva, K.K. (ed): *Yantrarāja of Mahendra Sūrī*, Bombay, 1936.

Sarma, Sreeramula Rajeswara (a): "Astronomical Instruments in Brahmagupta's Brāhmasphuṭasiddhānta," *Indian Historical Review*, New Delhi, 13 (1986-87) 63-74; reprinted in this Volume, pp. 19-46.

Sarma, Sreeramula Rajeswara (b): "Astronomical Instruments in Mughal Miniatures," *Studien zur Indologie und Iranistik*, Hamburg, 16-17 (1992) 235-276; reprinted in this Volume, pp. 76-121.

Sarma, Sreeramula Rajeswara (c): "Portable Instruments at Jaipur Observatory," read at the International Symposium on Indian & Other Asiatic Astronomies, Hyderabad-Jaipur, 1991.

Sarma, Sreeramula Rajeswara (d): "The Ṣafiḥa Zarqāliyya in India," in: *From Baghdad to Barcelona : Studies in the Islamic Exact Sciences in Honour of Prof. Juan Vernet*, Barcelona, 1996, pp. 719-735; reprinted in this Volume, pp. 223-39.

Sarma, Sreeramula Rajeswara (e): "Sulṭān, Sūri and the Astrolabe," *Indian Journal of History of Science*, 35.2 (2000) 129-147; reprinted in this Volume, pp. 179-98.

Sarma, S.R., Ansari, S.M.R & Kulkarni, A.G.: "Two Mughal Celestial Globes," *Indian J of History of Science*, New Delhi, 28.1 (1993) 80-89; reprinted in this Volume, pp. 294-307.

Seeman, Hugo J.: *Die Instrumente der Sternwarte zu Marāgha nach den Mitteilungen von al 'Urḍī*, Erlangen, 1929.

Seipel, Wilfried: *Mensch und Kosmos, Ober-Österreichische Landesausstellung*, 1990.

Sharma, R.S. (ed): *Samrāṭsiddhānta of Jagannātha*, New Delhi, 1967.

Smith, David Eugene: *History of Mathematics*, Boston, 1925.

15

Two Mughal Celestial Globes

Next to the astrolabe, the celestial globe (Latin: *globus coelestis;* Arabic: *al-Kura*) was the most widely used astronomical instrument in pre-modern times. Hipparchus is said to have used it in the second century BC. Ptolemy, in the second century AD, devotes a whole chapter of his *Almagest* to its construction and use.[1] But it reached its fullest development only in the Middle Ages both in Europe[2] and in the Islamic World.[3] While the astrolabe is a versatile observational and computational instrument, the celestial globe is primarily a teaching tool, on which the apparent motion of the heavens and the relative positions of the fixed stars can be easily demonstrated.

Though there are some slight variations in its construction, the most common type consists of a spherical globe with the ecliptic, equator, tropics, polar circles, ecliptic latitude circles and other circles drawn upon it. Moreover, a large number of star positions

Jointly with S.M.R. Ansari and A.G. Kulkarni. First published in: *Indian Journal of History of Science*, 28.1 (1993) 80-89.

1. Book VIII, Chapter 3: cf. "The Almagest by Ptolemy", tr. R. Catesby Taliefero, in: *Great Books of the Western World*, vol. 16, Chicago, 1952, pp. 261-263; *Ptolemy's Almagest,* translated and annotated by G. J. Toomer, New York, 1984, pp. 404-407. For a Sanskrit rendering of this chapter, see Jagannātha's *Samrāṭsiddhānta*, ed. R.S. Sharma, vol. II, New Delhi, 1967, pp. 781-783.
2. Cf. Zinner, Ernst, *Deutsche und Niederländische Astronomische Instrumente des 11-18. Jahrhunderts,* Munich 1979, pp. 168-178.
3. Cf. Wiedemann, E., s.v. al-Kura, *Encyclopaedia of Islam,* 2nd edn, vol. V, p. 397; David A. King, Astronomical Instrumentation in the Medieval Near East, in: idem, *Islamic Astronomical Instruments,* London, 1987, I, pp. 4, 18.

are also marked on the globe according to the coordinates given either by Ptolemy (for the year AD 138), or by Ulūgh Beg (for 841 AH/AD 1437-38), or as calculated by others, sometimes even by the globe-maker himself. Around specific clusters of stars are drawn constellation figures as conceived in Hellenistic mythology, or as modified later in the Islamic World. The *Almagest* describes 48 such constellations constituted by some 1018 fixed stars and enumerates the ecliptic coordinates and magnitudes of these stars.[4] Therefore, it was customary to draw 48 constellation figures and mark 1018 star positions, although there are globes with less number of stars.

At the two equatorial poles of the globe, holes are made for the axis to pass through, and this axis is pivoted inside a meridian ring, so that the globe rotates around the axis. The meridian ring is mounted perpendicularly upon a horizon ring which is supported by three or four legs. The axis can be so adjusted in the meridian ring that it is inclined to the plane of horizon by an angle equal to the terrestrial latitude of the place where the globe is to be used. Then the globe simulates the motion of the heavens above that place and the star positions on the globe correspond to those in the skies.

The manufacture of these globes was a highly specialized job, requiring great skill in metal work, engraving and calligraphy in addition to a good knowledge of astronomy. The surviving specimens, therefore, constitute a valuable part of a nation's scientific, technological and artistic heritage. The extant celestial globes of Islamic provenance were surveyed in an exemplary manner by Emilie Savage-Smith in her *Islamicate Celestial Globes: Their History, Construction, and Use*.[5] Unfortunately, she did not make a fresh survey of the specimens in India, but depended on file pictures and the scarce published notices with the result that there are lacunae in her account of the celestial globe in India.[6]

4. Cf. *Ptolemy's Almagest*, tr. Toomer, pp. 339-399.
5. Washington D.C. 1985. Henceforth cited as ESS. References are to the entry numbers of the globes surveyed, to the pages, or to the figures.
6. Interestingly enough, of the 50 signed globes described in ESS, as many as 26 are of Indian origin.

In spite of its popularity in the Islamic World, where it was used in every *madrasah* for teaching astronomy (*'Ilm al-Haya't*), the celestial globe appears to be a latecomer to India.[7] It is mentioned for the first time in India in connection with the Mughal emperor Humāyūn (reign 1530-1556) who was reputed to be an expert in the use of the astrolabe and the celestial globe. Under his royal patronage, astrolabes and celestial globes began to be manufactured in India. Abū 'l-Fazl states in his *Ā'īn-i Akbarī* that Maulānā Maqṣūd of Herat, a courtier of Humāyūn, manufactured excellent astrolabes and celestial globes,[8] but none of these survived. The earliest known celestial globe emanates from the reign of Humāyūn's son Akbar (reign 1556-1605). It was made by one 'Alī Kashmīrī, son of Lūqmān, in 998 AH/AD 1589-90. Today, the globe belongs to a private collection in London. Apart from this single creation, no other information is available on the life and work of 'Alī Kashmīrī.[9]

In this and the next century, several members of another family distinguished themselves as versatile and prolific manufacturers of astrolabes and celestial globes. The patriarch of this family, master Allāhdād of Lahore, was a near contemporary of Humāyūn, and is known for three astrolabes, one of which is dated 975 AH/AD 1567-68. But the earliest surviving celestial globes manufactured in this family are the four specimens made by his grandson Qā'im Muḥammad between the years AD 1622 and 1637.[10] Until then, celestial globes were made first as two concave hemispheres which were later joined at the line of equator. Qā'im Muḥammad perfected the technique of casting them as single hollow pieces by the *cire perdue* or lost wax process.[11] His son Ḍiyā' al-Dīn Muḥammad (*fl.* 1637-1680) excelled in this technique. Some 15 celestial globes crafted by him—the largest number ever known to have been made by a single individual—are preserved today in various parts of the

7. Cf. Sarma, Sreeramula Rajeswara, From al-Kura to Bhagola: On the Dissemination of the Celestial Globe in India, *Studies in History of Medicine and Science*, 13.1 (1994) 69-85; reprinted in this Volume, pp. 275-93.
8. *Ā'īn-i Akbarī*, tr. H. Blochmann, vol. I, New Delhi, 1977, pp. 54-55.
9. ESS, No. 10; pp. 34, 223-224; Figs. 11, 68.
10. ESS, Nos. 11-14; pp. 224-225.
11. ESS, pp. 34-44, 90-95.

world. He also made astrolabes of high artistic quality, of which more than 25 specimens exist in different collections.[12]

Ḍiyā' al-Dīn's contemporary, Muḥammad Ṣāliḥ of Thatta (*fl.* 1659-1667), was also a notable instrument-maker, though not so prolific as the former. So far, the following three astrolabes and two celestial globes made by him have been noticed.

ASTROLABES

(i) Dated 1076 AH/AD 1665-66, with a diameter of 124 mm, and deposited in the East India Company Museum, London; (ii) dated 1077 AH/AD 1666-67, with a diameter of 194 mm, and deposited in the Billmeir Collection of the Museum of the History of Science at Oxford; and (iii) in a private collection at Edgerton.[13]

CELESTIAL GLOBES

(i) The first one, measuring 210 mm in diameter, was manufactured in 1070 AH/AD 1659-60 by the order of one Shaykh 'Abd al-Khāliq. It is now preserved in the Archaeological Museum at the Red Fort, Delhi. Some 1018 star positions are marked on it with inlaid silver points, and a number of constellation figures are engraved in outline. There are holes both at the poles of the equator and those of the ecliptic so that the globe can be made to rotate around the equatorial axis as well as the ecliptic axis. The original stand is lost, and the globe is now mounted horizontally on a non-functional stand.[14]

(ii) The second known celestial globe by Muḥammad Ṣāliḥ was manufactured in 1074 AH/AD 1663-64, and is now in a private collection in London. This globe also exhibits the full complement of 1018 star positions marked by inlaid silver points and 48 constellation figures drawn in outline, besides the usual circles. It is mounted on a quadruped stand, with spindle-like legs, on a base formed by two crossbars. In the horizon ring there are two notches for the meridian ring to pass through. To its underside is attached

12. ESS, pp. 38-43; Sarma, op. cit.
13. ESS, p. 303, n. 158.
14. ESS, No. 25, pp. 229-230; B.L. Dhama, *Annual Report of the Archaeological Survey of India*, 28, 143-144, 1928-29.

a semicircular arc as a support for the meridian ring. The original meridian ring, however, is missing, and is substituted by an ungraduated modern ring.

But the most interesting aspect of this globe is that it is bilingual. It contains the names of the zodiac signs, lunar mansions, and other constellations in both Arabic (Naskh script) and Sanskrit (Devanāgarī script). The graduations are also numbered in two systems. The ecliptic is graduated in single degrees and each fifth is labelled in Arabic and Devanāgarī numerals. The equator is also graduated in single degrees and each fifth is labelled in *abjad* notation on one side and in Devanāgarī numerals on the other. In the case of the horizon ring, however, each sixth is labelled in Devanāgarī numerals only.[15]

Occasionally, the Sanskrit names are abbreviated, as there is not enough space for the full form. Thus, there are names of zodiac signs like *kuṃ. mū.* (for *kuṃbha-mūrti,* literal translation of *sūrat sākib al-mā'*, "figure of Aquarius'); *mī. mū.* (*mīna-mūrtī,* translation of *sūrat al-ḥut,* "figure of Pisces"); lunar mansions like *abhijit, śravaṇa, dhaniṣṭha, śatabhiṣā, pū. bhā.* (for *pūrvā bhādrapadā*) and constellations like *aśvaskandha* (*faras a'ẓam,* Pegasus); *matsya mū(rti) (al-dalfin,* Dolphin), etc.[16]

Near the south pole, there is a short inscription in Arabic to the effect that the globe was "the work of Muḥammad Ṣāliḥ Tatawī (in the) year 1074 (AH)." There is also a Sanskrit inscription, but Savage-Smith has not recorded the text. Even so, it is quite clear that Muḥammad Ṣāliḥ added the Sanskrit legends, so that the globe could be used not only by the Arabic-reading Muslim astronomers but also by those Hindu *jyotiṣīs*, who were interested in Islamic astronomy. Savage-Smith thinks that such clientele existed in the 17th century Kashmir and, for this reason, she wishes to assign Muḥammad Ṣāliḥ to Kashmir.[17] There is, however, no justification for this ascription. The Hindu *jyotiṣīs* who interacted with Islamic astronomy in the 16th and 17th centuries were mostly residents of Benares and / or were associated with the Mughal court at Agra,

15. ESS, No. 29, pp. 231-232.
16. ESS, Fig. 18 on p. 45.
17. ESS, p. 44.

Fatehpur Sikri or Delhi.[18] To cite a few instances, some 40 years before Muḥammad Ṣāliḥ crafted the bilingual globe, Nṛsiṃha Daivajña of Benares described at length how Muslims constructed the celestial globe and marked the positions of the fixed stars upon it.[19] His son Kamalākara shows a great familiarity with Islamic astronomy in his *Siddhāntatattvaviveka* (1658). Again, in 1643, one Mālajit, who received the title Vedāṅgarāya from Shāh Jahān, composed at Agra his *Pāraśīprakāśa*, which contains a glossary of Arabo-Persian technical terms pertaining to astronomy and astrology, explained through the medium of Sanskrit.[20]

It is for astronomers like these that Muḥammad Ṣāliḥ produced his bilingual globe. Whether he was born at Thatta in the Indus Delta or not, there cannot be much doubt about his place of activity being in the vicinity of the Mughal court at Agra or Delhi, where he played the role of an intermediary between the Islamic and Hindu traditions of Astronomy (see Postscript).

Therefore, other creations by this Mughal instrument maker also deserve attention. We are happy to announce in this paper the existence of two new Mughal celestial globes: one was signed by Muḥammad Ṣāliḥ and the other has also a fair chance of being his work. We saw these two globes in January 1992 in a private collection at Bangalore. These are said to have been acquired somewhere in Rajasthan in 1991. In the following report we shall call the signed globe A (Figs. 15.1, 15.2, 15.5) and the other one B (Figs. 15.3, 15.4, 15.6).

18. Cf. Pingree, David, Islamic Astronomy in Sanskrit, *Journal for the History of Arabic Science*, 2.2, 315-330, 1978; Sreeramula Rajeswara Sarma, Astronomical Instruments in Mughal Miniatures, *Studien zur Indologie und Iranistik*, 16-17, 235-76, 1992; reprinted in this Volume, pp. 76-121.
19. In his commentary (AD 1621) on Bhāskarācārya's *Siddhāntaśiromaṇi*, ed. Muralīdhara Caturveda, Varanasi, 1981, p. 438. Cf. also Sreeramula Rajeswara Sarma, "From al-Kura to Bhagola . . ." (n. 7 above).
20. Cf. Sarma, S.R., Islamic Calendar and Indian Eras, in: G. Kuppuram and K. Kumudamani (ed), *History of Science and Technology in India*, Delhi, 1990, vol. II, pp. 433-441.

GLOBE A

It is a hollow seamless globe, cast as one single piece by the *cire perdue* process, without any visible plugs. The diameter of the globe is 150 mm. On the globe, the celestial equator and the ecliptic are represented by double bands of lines. The narrower band is graduated in single degrees, while on the broader band groups of 6° are marked and numbered in *abjad* notation (see Fig. 15.2). The tropics, polar circles and those around the ecliptic poles are marked by single lines. The same is the case with the ecliptic latitude circles: these six great circles cut perpendicularly across the ecliptic and converge at its two poles. The solstitial and equinoctial colures are also represented by single lines. These two circles cut across the equator perpendicularly and converge at the two equatorial poles (see Fig. 15.1).

Fig. 15.1. Globe A with stand.

Fig. 15.2. Globe A with the horizon ring. Note the equator at the middle of the globe, Libra to its left and Serpentarius to the right.

In the other two globes by Muḥammad Ṣāliḥ and in the majority of the globes by his contemporary Ḍiyā' al-Dīn, the star positions are marked by inlaid silver points. But here they are indicated by engraved dots enclosed within small circles. This device appears also in two of Ḍiyā' al-Dīn's globes.[21] However, Ḍiyā' al-Dīn draws regular circles around the dots, whereas in the present globe the circles are irregular and appear to simulate the twinkling of the stars (see especially Figs. 15.5 and 15.6). The full complement of 1018 stars are represented in this manner. The outlines of all 48 constellation figures are executed meticulously. There is a dark finish on the surface of the globe. Against this background, the

21. ESS, Nos. 69 and 71, pp. 252-253. See also G.R. Kaye, *Astronomical Instruments in the Delhi Museum* (Memoirs of the Archaeological Survey of India, No. 12), Calcutta, 1921, pp. 16-19, Fig. 9.

engraved star points, constellation figures and circles stand out clearly. The maker's inscription is inside the southern polar circle and reads as follows: *'amal Muḥammad Ṣāliḥ Tatawī 1073*, "the work of Muḥammad Ṣāliḥ Tatawī[22] (made in the Hijrī year) 1073." This year corresponds to AD 1662-63.

The axis of the globe passes through the two equatorial poles and rests on the horizon ring. There are decorative end pieces at both sides of the axis. Like the equator and the ecliptic, the horizon ring also has two bands of lines. The narrower band is graduated in single degrees and the wider band in groups of 6° which are, however, not numbered (Fig. 15.2). The horizon ring is supported by a quadruped stand, having a height of 130 mm (Fig. 15.1). This stand resembles that of the bilingual globe just described. This one also has four spindle-like legs which stand on a base formed by two crossbars. However, there is a vital difference. There is no meridian ring and the axis of the globe is attached to the horizon ring. On the horizon ring, there are no notches to allow insertion of the meridian ring, nor is there a semicircular support below for the meridian ring. In this state, the axis of the globe cannot be adjusted for any local latitude save the terrestrial equator. It appears that the original horizon ring, the meridian ring and its support were either lost or damaged, and these were replaced later by a horizon ring which was graduated but not labelled and did not have any provision for the insertion of the meridian ring. The same is the situation with the stand of Globe B described below.

Globe B

The second globe is considerably larger. Its diameter measures 235 mm and the stand has a height of 210 mm (Figs. 15.3, 15.4). In all other respects, it closely resembles Globe A. The surface of this globe is much darker, almost black, and the engravings are much sharper. Unfortunately, however, this globe carries no maker's inscription. But a comparison of the two globes shows such exact

22. Savage-Smith, p. 44, draws attention to the divergence of the spelling of this word in the globes discussed above. While it is written as "Tatawī" in the bilingual globe of AH 1074, the word is spelled as "Tatah-wī" in the earlier one executed in AH 1070.

correspondence that the conclusion is inevitable that Globe B is also a creation of Muḥammad Ṣāliḥ. Both globes have solstitial and equinoctial colures, which is somewhat rare in Indian globes, and the iconography of the constellation figures is exactly the same. Compare, for example, the constellation Argo Navis (Arabic: *al-safīnah*) near the south pole in both the globes (Figs. 15.5 and 15.6). Though there is some slight divergence in star positions, the ships look quite alike, so do the bearded figureheads on the left with the jaunty turban.

With the discovery of these two new globes, Muḥammad Ṣāliḥ's known oeuvre has risen to seven pieces: three astrolabes and four celestial globes. It is hoped that a fresh survey may reveal more creations of this Mughal instrument-maker and throw further light on his life and work. While concluding this report, we should like to emphasize the urgent need for locating and documenting the scientific instruments manufactured in pre-modern times in India, before such instruments are lost for ever due to our neglect.[23]

23. Sarma, S.R. is currently cataloguing Indian astronomical instruments and welcomes information on extant specimens.

Fig. 15.3. Globe B with stand. Constellations of the northern hemisphere.

Fig. 15.4. Globe B with stand. Constellations of the southern hemisphere.

Fig. 15.5. Globe A, detail. Constellation Argo Navis.

Fig. 15.6. Globe B, detail. Constellation Argo Navis with Centaur to the right.

Postscript on Muḥammad Ṣāliḥ's "Bilingual" Globe of 1074 AH/AD 1663

This globe is now in the Nasser D. Khalili Collection of Islamic Art, London. I had the opportunity of studying this globe in July 2005 through the courtesy of Dr Nasser D. Khalili and Ms Nahla Nasser. The Sanskrit labels on this globe were not engraved by Ṣāliḥ as I thought earlier. These were added later on at the instance of Nandarāma in 1767. There is an inscription on the globe to this effect which reads as follows:

śuciśuddhasya pañcamyāṃ siddhanāgenduvatsare/
Nandarāmeṇa golo 'yaṃ kṛtaḥ sopaskaro mude//

"On the fifth day of the bright half of Āṣāḍha in the [Vikrama] year 1824, this [celestial] globe was endowed with additional [labels in Sanskrit] by Nandarāma for [his own] pleasure." The date translates to Wednesday 1 July 1767. Nandarāma (*fl.* 1763-1778) composed several works on Jyotiḥśāstra, including one on instruments with the title *Yantrasāra*. (See David Pingree, *Census of the Exact Sciences in Sanskrit*, A-3, Philadelphia, 1976, pp. 128-130). In the course of my survey, I came across about a dozen Arabic/Persian instruments on which Sanskrit labels were added later so that these instruments could be used by Hindu astronomers who did not read Arabic/Persian. Among such instruments, the present globe occupies a special place because it is the only one where we know the person responsible for the Sanskrit labels. The additions on the other instruments are anonymous. Nevertheless, these instruments, reworked with Sanskrit labels, testify to the Hindu astronomers' continuous interest in Islamic astrolabes and celestial globes.

Index

A Computerized Checklist of Astrolabes 23
Abd al-Rahīm Khān, Khān-i-Khānān 63, 106
Abjad notation 227, 230, 271, 298, 300
Abū 'l-Faiz Faizī 125
Abū 'l-Fazl 30, 77, 78, 89, 101, 102, 106, 276, 267, 296
  on Humāyūn and the celestial globe 277
  on Humāyūn's interest in astronomy 276-7
  on Jotik Rāi 102
  on Maulānā Chānd 77, 101
  on Maulānā Maqsūd 296
  on the institution of Ghariyālā 30
  on the time of Akbar's birth 77-8, 89
  participates in the translation of Ulūgh Beg's tables into Sanskrit 106
Adler Planetarium, Chicago 23, 25, 45
Ā'īn-i Akbarī (by Abū 'l-Fazl, 16th c.) 30, 89, 138, 278, 296
ajasra-yantra 64, 70
Ajmer, latitude of 188
Akbar 86, 94, 126, 276, 296
  Abū 'l-Fazl on the time of his birth 77-8
  his birth depicted in a miniature 84-5
  Hindu astrologers at his court 100-7
  time-keeping under 130, 138
Akbarnāma (by Abū 'l-Fazl, 16th c.) 77, 89, 102, 114
Akhlāq-i Naṣīrī 98
akṣapatra 267
al-Bīrūnī
  his various writings on the astrolabe 181-2, 184, 191
  manual on the astrolabe in Sanskrit verse 244
  may have introduced the astrolabe into India 37, 203
  on time-keeping establishments at Peshawar 129, 157
al-Hassan, Ahmad Y. 68, 71-2
'Alī Kashmīrī ibn Luqmān 43, 279, 296
alidade 180, 186, 202, 203, 226, 207, 266, 277
Aligarh globe 281-5
al-kura 92, 200, 275, 294
al-kursī 92
Allāhdād Asṭurlābī Humāyūnī Lāhūrī 38-40, 42, 44, 80, 204-5, 207-9, 223, 267, 278-9, 282, 286, 296
  astrolabes by 209-10, 218, 222
Almagest (by Ptolemy) 92, 200, 275, 291, 294, 295
al-Marrākushī 244
almucantars 180, 225, 269
alphabetical notation 257, 271
altitude(s) 226, 239
  sun's 33, 36, 77, 79, 203
  sun's meridian altitude 207, 208, 209, 266
  stellar 76

Alyās, Maulānā 102
Amber, latitude of 32
'ankabūt 180, 202, 267
Anup Sanskrit Library, Bikaner 247
Archaeological Museum, Golconda Fort, Hyderabad 135
Archaeological Museum, Red Fort, Delhi 24, 281, 288, 297
armillary sphere 61, 288
Arthaśāstra (by Kauṭilya) 28, 137
Āryabhaṭa I 23, 29, 47, 49, 54, 56, 58, 59, 70, 72, 94, 126, 127, 136, 137, 258, 276
  on the water clock 148-53
Āryabhaṭa II 155
Āryabhaṭasiddhānta (by Āryabhaṭa I, 5th c.) 29, 48, 58, 94, 148
Āryabhaṭīya (by Āryabhaṭa I, AD 499) 49, 56, 59, 143-4, 153, 162, 258
ascendant 36, 77, 79, 203, 234
astrolabe(s) 22, 23, 24, 25, 35, 36-42, 43, 45, 78, 79-84, 99, 134, 179-272, 275, 277, 278, 279, 281, 286, 288, 294, 296
  by Allāhdād 209-10, 218, 222
  by Ḍiyā' al-Dīn Muḥammad 212-13, 219-20, 224-5
  by Ḥāmid ibn Muḥammad Muqīm 220
  by 'Īsá ibn Allāhdād 209, 218
  by Jamāl al-Dīn ibn Muḥammad Muqīm 221
  by Muḥammad Muqīm 209-12, 218-19, 222
  by Muḥammad Ṣālih of Thatta 297
  by Qāim Muḥammad 218
  components of 180, 186
  composite 184
  depicted in Mughal miniatures 83-4
  Greek 78
  in Islamic culture 79-84
  Indo-Persian 40, 45, 207, 209, 242, 252, 260
  Islamic 271, 307
  manufacture of 80, 194, 203, 278
  Mughal 260, 266, 267, 269, 271
  nisfī 80
  northern 184, 193, 202, 224, 235, 244, 245
  north-south 183, 184, 185, 188, 193, 194, 212, 224, 225
  Persian 207
  Sanskrit 40, 41, 216, 240-72, 253
  Sanskrit manuals on 40, 81-3, 234, 242, 256, 278
  single-plate 42, 252, 260
  southern 184, 185, 245
  universal 39, 235, 280
  Zarqālī 213, 223-39, 248, 280
astrolabe-maker 263
astrolabe-makers of Lahore 24
astrolabist family of Lahore 38-40, 44, 92, 199-222, 235, 279
  patrons of 214
astrologers 44, 77, 78, 79, 84, 125
  Hindu astrologers at Akbar's court 100-7
  Mughal portraits of 98-100
astronomer(s) 79
  European portraits of 275
  Hindu 81, 83, 92, 182, 235, 275, 289, 309
  Jaina 81, 83, 182
  Mughal portraits of 98-100
  Muslim 235, 241, 288, 298
astronomical tables 77, 106, 263
Astronomisch-Physikalisches Kabinett, Kassel 23
asṭurlāb 180, 201
  asṭurlāb janūbī 184, 245
  asṭurlāb shumālī 184, 202, 144
  asṭurlāb shumālī wa janūbī 184, 193, 203, 245
  Asṭurlāb-i Fīrūz-Shāhī 184-9, 203, 225

Aurangzeb 24, 39, 42, 194, 199, 213, 223, 226, 286
automatic devices 56, 59, 62
azimuth lines 180, 202, 225, 239, 269

Bābur 30, 85, 138
Bahlūmal Lāhūrī, Lālah 40, 43, 44, 215, 288, 291; *see also* Buhlomalla; Buhlovarma
Baloch, N.A. 133
Bāpudeva Śāstrin 250, 256
Bauernring 22, 97
Bhagola, bhagola-yantra 91, 275, 289
bhapatra 267
Bharat Kala Bhavan 25
Bhāskara I
  on the method of levelling the ground 49-50
  on measuring time with long syllables 143-6, 162-3
  on the water clock 153-4, 156, 157
Bhāskara II 20, 48, 92, 105, 126, 289
  attitude towards instruments 55
  legend of his daughter Līlāvatī 125
  on perpetual motion wheels 61, 65-74
  on the water clock 136, 151-3, 155
  rejects automatic devices 60
Bhoja 75
Bījagaṇita (by Bhāskara II, 12th c.) 106
Bist Bāb (by Naṣīr al-Dīn al-Ṭūsī, 13th c.) 195, 206, 249; *see also* *Risālah-i Bist Bāb dar Ma'rifat-i Usṭurlāb*
Book of Ingenious Devices (by al-Jazarī, *ca* 1205) 21
Brahmagupta 34, 73, 152
  on instruments 20, 47-63
  on perpetual motion wheels 70, 72
Brāhmasphuṭasiddhānta (by Brahmagupta, 628) 20, 47-63, 70, 72
Brieux, Alain 199
British Library, London 114
British Museum, London 26
Buddhaghoṣa 127, 136, 140
Buhlomalla 44; *see also* Bahlūmal Lāhūrī; Buhlovarma
Buhlovarma, Lalah 44; *see also* Bahlūmal Lāhūrī; Buhlomalla
Bundi, latitude of 255

cābuka-yantra 33
cakra, cakra-yantra 20, 51, 52, 53, 94
cardinal direction(s) 72, 159, 162
cardinal points, Indian method of determining 21
celestial globe(s) 23, 24, 36, 37, 39, 40, 42-5, 91-3, 99, 116, 199, 200-1, 204, 205, 214, 215, 223, 275-307
  by Ḍiyā' al-Dīn Muḥammad 221, 279-85
  by Ghulām Ḥusain Jaunpūrī 45, 289
  by Ḥāmid ibn Muḥammad Muqīm 221-2
  by Muḥammad Muqīm 221
  by Muḥammad Ṣālih of Thatta 297-307
  by Qāim Muḥammad 205, 213, 221-2, 279
  depicted in Mughal miniatures 91-3
  Humāyūn's interest in 277
  Islamic 307
  Mughal 294-307
  Sanskrit 289-91
Centaurus 282

Chānd, Maulānā 77, 78, 80, 101, 102
Chaucer, Geoffrey 243
Chester Beatty Collection, Dublin 113
Christie 26, 235
cire perdue 205, 213, 281, 288, 296, 300
City Palace Museum, Jaipur 25
clepsydra 55, 59
outflow type 58, 59, 62
sinking bowl type 51, 56
climate(s) 183, 184, 188
Columbia University, New York 25, 45, 290
column dial
of damascened steel 33
of ivory 33
wooden 33, 34, 45
Copernicus 275
cūḍā-yantra 22, 23, 93, 96, 97, 98, 100, 121
Cūḍāyantrasya Sāraṇī 96

Damodara 40, 251
Delhi, latitude of 22, 32, 96, 192
Dhanuṣ 20, 51, 52, 53
Dharmasindhu (by Kāśīnātha Upādhye, 1790-91) 156, 158, 160, 162, 171-4
dhāt al-Kursī 275
dhī-yantra 54
dhruvabhrama-yantra 27, 35, 44, 134
diksādhanā 21
Ḍiyā' al-Dīn Muḥammad 39, 42-4, 199, 206-7, 212-15, 223-39, 248, 266, 281-2, 286, 288-9, 296, 301
astrolabes by 212-13, 219-20, 224-5
celestial globes by 221, 223-4, 279-85
north-south astrolabe by 224-5
Zarqālī astrolabe by 223, 225-30

equinoctial dial, adjustable 45
engraved in Persian 44

Faizī Sarhindī 89
Faridul Emadi 33
Fathullāh Shīrāzī 102, 106
Firūz Shāh Tughluq
astrolabe manufacture under 37, 182-98, 203, 225, 241, 278
Asṭurlāb-i Fīrūz-Shāhī 184-9, 203, 225
interest in the astrolabe 79-80
sets up a water clock at his court 30, 85, 130, 138
sponsors a Sanskrit manual on the astrolabe 21, 40, 81, 225, 241, 243, 244, 278
Forbes, Eric 76

Garrett (A. ff) 24, 227
geographical gazetteer 187-8, 191, 202, 207, 208, 261, 263-4
ghaḍiyāl 130
ghaḍiyāla 30, 148
ghariyāl 130
ghariyālīs 30, 130, 138
ghaṭī-gṛha 129
ghaṭikāgāra 134
ghaṭikā-gṛha 129
ghaṭikā-gṛha-karaṇa 129
ghaṭikālaya 129, 130, 148
ghaṭikā-yantra 29, 84-6, 94, 100, 126, 137, 146-8, 154, 160, 163
Ghaṭikāyantraghaṭanāvidhi (anon, *ca.* 17th c.) 144, 158-68
ghaṭī-yantra 126, 130, 137, 155
Ghulām Ḥussain Jaunpūrī
celestial globe by 45, 289
Gibbs, Sharon 263

Gilchrist (John) 146
globus coelestis 294
gnomon 28, 31, 32, 51, 72, 186, 232, 246, 266
  horizontal 33
  triangular 32
  vertical 32
golden minaret 197, 198
Gopīrāja 243
Government Museum, Madras 29, 135
Govinda Daivajña 104-5, 158

Ḥāmid ibn Muḥammad Muqīm 44, 206, 213, 279
  astrolabes by 220
  celestial globes by 221-2
Harvard University, Cambridge 25
Havart, Daniel, on time-keeping 139-41
Hawa Mahal Museum, Jaipur 25, 289
Hayatnagar 139
Hemavijaya Gaṇī 146, 159
Hill, Donald R. 68, 71-2
Hipparchus 179, 201, 240, 294
Horniman Museum, London 34, 135
hours, equal 37, 79, 202
  seasonal 37, 202, 269
  unequal 79, 269
Humāyūn 24, 199
  anecdote concerning the celestial globe 277
  interest in astronomy 276-8
  may have introduced the celestial globe into India 42, 92
  patronage of astrolabists 37-8, 80, 209, 278, 296
  portrait of 93-4, 286

Ibn al-Zarqālluh 39, 213, 225, 230, 234-5, 248, 280
Ibrahīm ibn Said 276
Indian circle 21, 72
Indrajī 41, 252
'Īsá ibn Allāhdād 38, 204, 205, 208, 209; *see also* Mullā 'Īsá
  astrolabes by 209, 218
Islamic astronomical instruments 23, 45, 94; *see also* astrolabe, celestial globe
Islamic Technology: An Illustrated History (by Ahmad Y. al-Hasan & Donald R. Hill) 68
Islamic World 22, 33, 36, 47, 61, 65, 66, 71, 72, 73, 75, 88, 92, 93, 98, 134, 179, 200, 241, 242, 249, 275, 276, 294, 295, 296
Islamicate Celestial Globes, their History, Construction, and Use (by Emilie Savage-Smith) 23, 250, 295
Īśvaradāsa 103
I-Tsing 73, 129, 137, 157

Jābir al-Ṣūfī 37
Jahāngīr 63, 76, 90, 100, 101, 102, 103, 105, 205, 214, 251, 260, 279
Jai Singh, Sawāī 19, 20, 22, 23, 56, 100, 121, 215, 289
  and *cūḍā-yantra* 96
  and Zarqālī astrolabe 39-40, 226, 230, 232-6
  astrolabe production under 41-2, 252
  collected Mughal astrolabes 214
  on the astrolabe 247-9
  sponsored the composition and translation of books on instruments 82-3, 291
Jamāl al-Dīn ibn Muḥammad Muqīm 44, 206, 207, 213
  astrolabes by 221

Jāmi'-i Bahādur Khānī (by Ghulām Ḥussain Jaunpūrī, 1835) 289
Jātakapaddhati 106
Jemangala 34
Jñānarāja 88
Jotik Rāi 102-5
Jyautiṣa-saukhya (by Nīlakaṇṭha, 1572) 104
Jyotiṣarāya 102, 103

kamāl 76
Kamalākara 299
kapāla, kapāla-yantra 20, 51, 53, 148
kartarī 20, 51, 53, 54
kaśā-yantra 33
Kāśīnātha Upādhye 158
Kaṭapayādi notation 257-9, 269-72
Kathāratnākara (by Hemavijaya Gaṇī, 1600) 146, 159, 162, 163
Kaye (G.R.) 24, 250
Kennedy, E.S. 181
Kepler 275
Keśava 103
Khaṇḍakhādyaka (by Brahmagupta, 665) 47
Khanqah Emadia, Patna 33
Khuda Bakhsh Oriental Public Library, Patna 25, 36, 44, 183, 279
King, David 23, 199
Kṛṣṇa Daivajña 104-6
Kublai Khān 276

Lahore astrolabes, main features of 207-13
Lāl Shahbāz Qalandar 30, 133
Lalla 20, 48, 53, 59, 61, 136
  on the water clock 151-2
latitudes 49, 180, 208, 226, 232, 260, 262
Lekhapaddhati 129
Līlānātha Jyotirvid 251
Līlāvatī (by Bhāskara II, 12th c.) 125
Linton, Leonard 26
longitudes 208, 262
lunar mansions 209, 230, 262

Maddison, Francis 199
Madho Singh 20
Mahendra Sūri 21, 40, 81, 190-4, 243-4, 246, 248, 251, 255, 256, 278
Malayendu Sūri 81, 190-4, 243
Maṇirāma 40, 251
Maqsūd, Maulānā, of Herat 278, 296
Marāgha 100
Marīci commentary on the *Siddhāntaśiromaṇi* (by Munīśvara, before 1638) 153
Marshall, John 89
masonry instruments 23, 96, 230
mater 180, 186, 202, 261, 262
Mathurānātha Śukla 83, 249, 256
Megharatna Muni 82, 246, 256
miśra-yantra 194, 245
Morley, W.H. 250
Mu'ayyad al-Din al-'Urḍī 21, 72, 276
mudrikā-yantra 96
Mughal miniatures 22-3, 30, 76-121
Muḥammad bin Tughluq Shāh 182
Muḥammad Faḍlullāh of Aurangabad 45, 215
Muḥammad ibn Mu'ayyad al-Urḍī 276
Muḥammad Muqīm 39, 44, 205-6, 210-12, 223, 266, 279
  astrolabes by 209-12, 218-19, 222
  celestial globe by 221
Muḥammad Mūsá Asṭurlābī 45
Muḥammad Ṣālih of Thatta 43, 214-15, 286, 288, 297-307
  astrolabes by 297
  celestial globes by 297-307
Muhūrtacintāmaṇi (by Rāma Daivajña, 1600) 158

Mullā 'Īsá 39, 44, 282; *see also* 'Īsá ibn Allāhdād
Munīśvara 105, 153
Murād 76
Musée Guimet, Paris 286
Museum of Indology, Jaipur 25, 135, 290
Museum of the History of Science, Oxford 23, 26, 33, 204, 254, 279, 297
Museum of the Worshipful Company of Clockmakers, London 26, 135

nāḍikā-yantra 85
nāḍīvalaya-yantra 32
Nadvi, Syed Sulaiman 24, 80, 92, 199, 278, 282
nālikā-yantra 28
Nandarāma 83, 249, 256, 307
Nāradasaṃhitā 158, 160, 161
  on the water clock 168-70
Naṣīr al-Dīn al-Ṭūsī 82, 195, 206, 249, 291
Naṣīr al-Dīn Ḥasan 36
Nasser D. Khalili Collection of Islamic Art, London 307
National Maritime Museum, Greenwich 26, 34
Nawāb Abūl Ḥasan 205, 214
Nawāb Iftikār Khān of Jaunpur 39, 214, 226
Nawāb 'Itiqād Khān 214
Needham, Joseph 69, 71
Nīlakaṇṭha (Jotik Rāi) 104-5
Nīlakaṇṭha Somasutvan 55, 57, 59
Nityānanda 82, 247, 252
nocturnal 35
North, John 179
Nṛsiṃha Daivajña
  on the astrolabe 247, 256
  on the celestial globe 92, 289, 291, 299

observatory, Jai Singh's, at Jaipur 22, 23, 25, 33, 35, 96, 121, 224, 226, 252, 280
observatory, Marāgha 21, 72
observatory, Taqī al-Dīn's, at Istanbul 88

Padmanābha 35, 81, 245, 256
palabhā-yantra 32
pala-vṛtta(s) 146, 159, 162, 163
Pañcasiddhāntikā (by Varāhamihira, 6th c.) 48, 52, 96
Parameśvara 57, 144
  eclipse observations by 55
  on the astrolabe 246
  on the method of levelling the ground 50
  on the pair of compasses 50
Pārasīprakāśa (by Vedāṅgarāya, 1643) 299
perpetual motion machines 20, 64-75, 68, 73
perpetual motion wheel 65-8, 71, 75
perpetuum mobile 20, 60-1, 63-4, 66, 69-70, 72, 75
Peuerbach, Georg 97
phalaka-yantra 27, 55
Pingree, David 41
pīṭha 20, 51, 53
Pitt Rivers Museum, Oxford 26, 31, 41, 135, 252
Piyūṣadhārā (by Govinda Daivajña, 1603) 158, 161, 162
  on the water clock 170-1
pratoda-yantra 33
precession 192
Price, Derek 23, 199
Ptolemy 92, 200, 275, 294, 295

Qāim Muḥammad 39, 42, 44, 205, 206, 213, 214, 223, 281, 282, 288, 296
  astrolabes by 218

celestial globes by 205, 213, 221, 222, 279
quadrant(s) 34, 52, 288
double 35
Gunter's 260
horary 36
Indo-Persian 34
Sanskrit 35
wooden, Turkish style 35
quicksilver wheel 71

Rāghavajit 251, 252, 253
Rāma Daivajña 158
Rāmacandra Vājapeyin 21, 22, 23, 81, 90, 245, 247, 256
on the ring dial 95-6
on the sand clock 86-8
Rāmakṛṣṇa Ārādhya 48
Raṅganātha 63, 74, 105
Regiomontan 97
rete 180, 184, 186, 202-3, 207-8, 267
zoomorphic 212
ring dial 22-3, 77, 79, 84, 91, 93-8, 116
depicted in Mughal miniatures 93-4
European 98
Sanskrit, universal 44
Risālah-i Bist Bāb dar Ma'rifat-i Usṭurlāb (by Naṣīr al-Dīn al-Ṭūsī, 13th c.) 82; *see also Bist Bāb*
Royal Scottish Museum, Edinburgh 26, 41, 251
rūznumah 36

Sadratnamālā (by Śaṅkara Varman, 1819) 257
ṣafīḥa 202, 226, 267
Ṣafīḥa Zarqāliyya 223-39
śakaṭa 54
Salar Jung Museum, Hyderabad 25, 38, 212, 279
Saliba, George 263
Salīm (later Jahāngīr) 100, 106, 113
Samarqand 100, 208, 209
latitude of 35
sand clock 77, 84, 86-91, 99
depicted in Mughal miniatures 90-1
European 89-91
Śaṅkara Varman 257
śaṅku 20, 28, 51
Sanskrit University, Varanasi 25, 159, 249, 257-72
Sanskrit astronomical instruments (*yantra*)
ajasra-yantra 64, 70
bhagola, bhagola-yantra 91, 275, 289
cābuka-yantra 33
cakra, cakra-yantra 20, 51, 52, 53, 94
cūḍā-yantra 22, 23, 93, 96, 97, 98, 100, 121
dhanuṣ 20, 51, 52, 53
dhī-yantra 54
dhruvabhrama-yantra 27, 35, 44, 134
ghaṭikā-yantra 29, 84-6, 94, 100, 126, 137, 146-8, 154, 160, 163
ghaṭī-yantra 126, 130, 137, 155
kapāla, kapāla-yantra 20, 51, 53, 148
kartarī 20, 51, 53, 54
kaśā-yantra 33
miśra-yantra 194, 245
mudrikā-yantra 96
nāḍikā-yantra 85
nāḍīvalaya-yantra 32
nālikā-yantra 28
palabhā-yantra 32
phalaka-yantra 27, 55
pīṭha 20, 51, 53
pratoda-yantra 33
śakaṭa 54
śaṅku 20, 28, 51

Sarvadeśīya-Jarakālī-yantra 83, 232, 235, 238-9, 248, 256
saumya-yantra 193, 244
svayamṿaha-yantra 56, 70, 276
turīya-yantra 34, 36, 44
turyagola 50, 51, 52, 53
valaya, valaya-yantra 22, 95, 96
yāma-yantra 194
yāmya-yantra 245
Yantra jarakālī sarvadeśī 39, 232
yantrarāja 40, 79, 240-72
yaṣṭi 20, 51, 54
Sarasvati Bhavan Library, Varanasi 159, 249, 260
Sarvadeśīya-Jarakālī-yantra (anon, 18th c.) 83, 232, 235, 238-9, 248, 256
Sarvasiddhāntarāja (by Nityānanda, 1639) 247, 256
saumya-yantra 193, 244
Savage-Smith, Emilie 23, 199, 250, 279, 295, 281, 290, 298
Science Museum, London 26, 43, 252, 255
self-propelling machines 56, 63, 70
Sen, S.N. 24
shabnumah 36
shadow squares 186, 208, 246, 266
Shāh Jahān 39, 93, 94, 194, 279, 286, 299
Shams-i-Sirāj 'Afif 30, 79, 130, 183, 197
Shaykh 'Abd al-Khāliq 288, 297
Siddhāntadīpikā on the *Mahābhāskarīya* (by Parameśvara, 1432) 246, 256
Siddhāntarāja (by Nityānanda, 1639) 82
Siddhāntaśekhara (by Śrīpati, 1039) 48, 155
Siddhāntaśiromaṇi (by Bhāskara II, 1150) 48, 62, 65, 69, 73, 92, 151, 247, 289
Siddhāntatattvaviveka (by Kamalākara, 1658) 299
Siddhāntasundara (by Jñānarāja, 1503) 88
sine graph 208, 230
Sīrat-i Fīrūz-Shāhī (anon, 1370) 37, 183-98, 203, 244, 278
Śiṣyadhīvṛddhida (by Lalla, 8-9th c.) 48
Śivalāla 252, 255
Smith, David Eugene 290
Smithsonian Institution, Washington D.C. 25, 26, 45, 286
Someśvara 57
Sotheby 26, 40
Śrīdhara 151
Śrīnātha Chagāṇī 83, 249, 256
Śrīpati 20, 48, 53, 59, 106
  on the water clock 155
star catalogue 193
star map 180, 196, 267
State Museum of Archaeology, Hyderabad 25
stereographic projection 179, 181, 184, 187, 201-2
Su Sung's clock tower 62, 63, 69, 73
sundial 31, 32, 129
Sūryadeva Yajvan 57, 144, 258
Sūryasiddhānta 48, 62, 63, 69
svayamṿaha-yantra 56, 70, 276

Taqī al-Dīn 88
Tārīkh-i Fīrūz Shāhī (by Shams-i-Sirāj 'Afif, ca 1388) 183, 190, 197
ṭās-i ghariyāla 30, 85, 130, 138
Travernier (Jean Baptiste) 139
Time Museum, Rockford 25, 33
time, cyclical 71
  measurable 71
  sidereal 36
time-keeping 130
  establishments, institutions for 127, 129, 137, 139, 157

Timūr 76, 114
Timūrnāma 77
Treatise on the Astrolabe (by Geoffrey Chaucer, 14th c.) 243
Tropic of Cancer 184, 188
Tropic of Capricorn 184, 188, 245, 267
turīya-yantra 34, 36, 44
Turner, A.J. 214, 225
turyagola 50, 51, 52, 53
tympan(s) 180, 184, 186, 187, 202, 208, 251, 260, 267

Ulūgh Beg 98, 106, 200, 295, 281
Usturalāvayantra (by Megharatna Muni, 15th c.) 82, 246, 256

valaya, valaya-yantra 22, 95, 96
Vaṃśavalīnagara 268, 270, 271, 272
Varāhamihira 22, 48, 52, 54, 95, 136
Vāsanābhāṣya (by Bhāskara II on his own *Siddhāntaśiromaṇi,* 1150) 152
Vāsanāvārttika on the *Siddhāntaśiromaṇi* (by Nṛsiṃha Daivajña, 1621) 247, 256
Vedāṅgajyotiṣa 28
Vedāṅgarāya 299
Victoria & Albert Museum, London 26, 224
Viśrāma 82, 126, 247, 256
on the water clock 127

water clock(s) 28, 30, 73; 77, 84-6, 125-75
Āryabhaṭa I on 148-53
Bhāskara I on 153-4
Bhāskara II on 151-3
depicted in Mughal miniatures 84-5
Lalla on 150-2
made of coconut shell 29
outflow type 28, 52, 70, 149, 154, 156, 161, 276
sinking bowl type 28, 45, 84, 88, 126, 154, 156, 161
working principle 126
water clock house 129, 130, 134
Welch, Stuart Cary 100
Welsh National Museum, Cardiff 26, 224
Whipple Museum of History of Science, Cambridge 26, 279
White, Lynn 20, 61, 64-72, 74, 75
Wilkinson, Lancelot 69, 73
Wilson, H.H. 250

Yajñeśvara 243
yāma-yantra 194
yāmya-yantra 245
yantra, see Sanskrit astronomical instruments
Yantra jarakālī sarvadeśī 39, 232
Yantrakiraṇāvalī (by Padmanābha, *ca.* 1400) 245, 256
Yantraprabhā (by Śrīnātha Chagāṇī, 18th c.) 83, 249, 256
Yantraprakāra (of Sawāī Jai Singh, 18th c.) 22, 32, 82, 96, 248, 256
Yantraprakāśa (by Rāmacandra Vājapeyin, 1428) 21, 22, 33, 81, 86, 95, 245, 256
Yantrarāja (by Mahendra Sūri, 1370) 21, 40, 190-4, 225, 243-5, 252, 255, 256
yantrarāja, Sanskrit astrolabe 40, 79, 240-72
Yantrarājaghaṭanā (by Mathurānātha Śukla, 1782) 83, 249, 256
Yantrarājaracanā (by Sawāī Jai Singh 18th c.) 82, 83, 235, 248, 249, 256
Yantrarājavicāraviṃśādhyāyī (anon, 18th c.) 83, 249, 256

Yantrarājopayogi-chedyaka (by Bāpudeva Śāstrin, 19th c.) 250, 256
Yantrarājādhikāra (by Padmanābha, *ca* 1400) 245
Yantrasāra (by Nandarāma, 1772) 83, 249, 256, 307
Yantraśiromaṇi (by Viśrāma, 1615) 82, 126, 247, 256
yaṣṭi 20, 51, 54

Zarqālī astrolabe 213, 223-39, 248, 280
Zayn al-'Ābidin of Delhi 215, 288
Zinner, Ernst 23
zodiac figures 98
zodiac signs 99, 207, 208, 209, 227, 230, 266, 267, 268, 298